Lecture Notes in Computer Science 10248

Commenced Publication in 1973
Founding and Former Series Editors:
Gerhard Goos, Juris Hartmanis, and Jan van Leeuwen

More information about this series at http://www.springer.com/series/7407

Alberto Dennunzio · Enrico Formenti
Luca Manzoni · Antonio E. Porreca (Eds.)

Cellular Automata and Discrete Complex Systems

23rd IFIP WG 1.5 International Workshop, AUTOMATA 2017
Milan, Italy, June 7–9, 2017
Proceedings

 Springer

Editors
Alberto Dennunzio
Dipartimento di Informatica,
  Sistemistica e Comunicazione
Università degli Studi di Milano-Bicocca
Milano
Italy

Enrico Formenti
CNRS, I3S
Université Côte d'Azur
Nice
France

Luca Manzoni
Dipartimento di Informatica,
  Sistemistica e Comunicazione
Università degli Studi di Milano-Bicocca
Milano
Italy

Antonio E. Porreca
Dipartimento di Informatica,
  Sistemistica e Comunicazione
Università degli Studi di Milano-Bicocca
Milano
Italy

ISSN 0302-9743   ISSN 1611-3349 (electronic)
Lecture Notes in Computer Science
ISBN 978-3-319-58630-4   ISBN 978-3-319-58631-1 (eBook)
DOI 10.1007/978-3-319-58631-1

Library of Congress Control Number: 2017940244

LNCS Sublibrary: SL1 – Theoretical Computer Science and General Issues

Printed on acid-free paper

This Springer imprint is published by Springer Nature
The registered company is Springer International Publishing AG
The registered company address is: Gewerbestrasse 11, 6330 Cham, Switzerland

Preface

The 23rd International Workshop on Cellular Automata and Discrete Complex Systems, AUTOMATA 2017, was held in Milan, Italy, during June 7–9, 2017.

It was organized by the Department of Informatics, Systems, and Communication of the University of Milano-Bicocca. The event was an IFIP Working Conference and it hosted the annual meeting of the IFIP Working Group 1.5.

AUTOMATA 2017 continued an annual series of events established in 1995 as a forum for the collaboration between researchers in the field of cellular automata and related discrete complex systems. Over the years, the topics have been progressively expanded. This year the scope was further broadened including new topics concerning correlated models of automata.

Current topics include (but are not limited to) the following aspects and features of such systems: dynamics, topological, ergodic, and algebraic aspects, algorithmic and complexity issues, emergent properties, formal languages, symbolic dynamics, tilings, models of parallelism and distributed systems, timing schemes, synchronous versus asynchronous models, phenomenological descriptions, scientific modelling, and practical applications. The conference attracted a good number of submissions, which indicates a continued interest in these topics.

There were three invited talks at the conference, and we wish to thank the speakers Eric Goles, Adrien Richard, and Ville Salo for accepting the invitation and for their very interesting presentations. The invited contributions are included in this volume.

There were 29 submissions as full papers to the conference. Each submission was managed by two or three Program Committee members. Based on the reviews and discussions, the committee decided to accept 14 papers to be presented at the conference and to be included in the proceedings. We would like to thank all authors for their contributions and work without which this event would not have been possible. The conference program also involved short presentations of exploratory papers that are not included in these proceedings, and we wish to extend our thanks also to the authors of the exploratory submissions.

We are indebted to the Program Committee and the additional reviewers for their valuable help in selecting the papers. We extend our thanks to the remaining member of the local Organizing Committee, Luca Mariot. We are also grateful for the support by the Department of Informatics, Systems and Communication and the University of Milano-Bicocca. Finally, we acknowledge the excellent cooperation from the *Lecture Notes in Computer Science* team of Springer for their help in producing this volume in time for the conference.

April 2017

Alberto Dennunzio
Enrico Formenti
Luca Manzoni
Antonio E. Porreca

Organization

Program Committee

Elena Barcucci	Università degli Studi di Firenze, Italy
Alberto Dennunzio	Università degli Studi di Milano-Bicocca, Italy, Co-chair
Bruno Durand	Université Montpellier 2, France
Nazim Fatès	Inria Nancy, France
Paola Flocchini	University of Ottawa, Canada
Enrico Formenti	Université Côte d'Azur, France, Co-chair
Anahí Gajardo	Universidad de Concepción, Chile
Dora Giammarresi	Università degli Studi di Roma Tor Vergata, Italy
Eric Goles	Universidad Adolfo Ibáñez, Chile
Katsunobu Imai	Hiroshima University, Japan
Jarkko Kari	University of Turku, Finland
Petr Kůrka	Charles University in Prague, Czech Republic
Martin Kutrib	Justus-Liebig-Universität Gießen, Germany
Andreas Malcher	Justus-Liebig-Universität Gießen, Germany
Carlos Martín-Vide	Universitat Rovira i Virgili, Spain
Giancarlo Mauri	Università degli Studi di Milano-Bicocca, Italy
Kenichi Morita	Hiroshima University, Japan
Pedro de Oliveira	Universidade Presbiteriana Mackenzie, Brazil
Ronnie Pavlov	University of Denver, USA
Karl Petersen	University of North Carolina at Chapel Hill, USA
Ion Petre	University of Helsinki, Finland
Renzo Pinzani	Università degli Studi di Firenze, Italy
Siamak Taati	University of British Columbia, Canada
Edgardo Ugalde	Universidad Autónoma de San Luis Potosí, Mexico
Hiroshi Umeo	Osaka Electro-Communication University, Japan

Steering Committee

Teijiro Isokawa	University of Hyogo, Japan
Jarkko Kari	University of Turku, Finland
Turlough Neary	Universität Zürich and ETH Zürich, Switzerland
Pedro de Oliveira	Universidade Presbiteriana Mackenzie, Brazil
Thomas Worsch	Universität Karlsruhe, Germany

Organizing Committee

Alberto Dennunzio	Università degli Studi di Milano-Bicocca, Co-chair
Enrico Formenti	Université Côte d'Azur, Co-chair
Luca Manzoni	Università degli Studi di Milano-Bicocca

Luca Mariot   Università degli Studi di Milano-Bicocca
                          and Université Côte d'Azur
Antonio E. Porreca   Università degli Studi di Milano-Bicocca

Additional Reviewers

Mikhail Barash   Florin Manea
Silvio Capobianco   Luca Manzoni
Michel Coornaert   Luca Mariot
Julien Destombes   Alexander Okhotin
Cinzia Di Giusto   Bartłomiej Płaczek
Pietro Di Lena   Antonio E. Porreca
Francesca Fiorenzi   Victor Poupet
Anaël Grandjean   Felipe Ramos
Joonatan Jalonen   Klaus Sutner
Johan Kopra   Ilkka Törmä

Sponsoring Institutions

 Università degli Studi di Milano-Bicocca

 Dipartimento di Informatica, Sistemistica e Comunicazione
Università degli Studi di Milano-Bicocca

Abstracts of Invited Talks

Two Dimensional Cellular Automata and Computational Complexity

Eric Goles

Facultad de Tecnología y Ciencias, Universidad Adolfo Ibáñez, Santiago, Chile

Let us consider a two-dimensional automata with states $\{0, 1\}$ and the von Neumann neighborhood. I will present results about the computational complexity of some prediction problems related with the strict majority as well as the class of freezing local functions.

The strict majority function considers the most represented state in its neighborhood. In case of tie the central state remains unchanged. On the other hand, in a freezing local functions the state 1 remains invariant, so dynamics appears only by updating sites in state 0. For the von Neumann neighborhood (considering 0 as a quiescent state) there are 32 rotation invariant local functions, so 16 totalistic ones, i.e., depending only of the sum of the states). In the next table we exhibit the totalistic local rules:

Table 1. Possible rules and their complexity when $0 \notin I$. The ✓ in the i-th column means that the local rule is 1 for that value of the sum.

Rule	$1 \in I$	$2 \in I$	$3 \in I$	$4 \in I$	EventPred	AsyncPred
T					$\mathcal{O}(1)$	$\mathcal{O}(1)$
4				✓	$\mathcal{O}(1)$	$\mathcal{O}(1)$
3			✓		in NC	?
34			✓	✓	in NC	NC
2		✓			P-Complete	?
24		✓		✓	P-Complete	?
23		✓	✓		?	?
234		✓	✓	✓	in NC	NC
1	✓				?	in NC
14	✓			✓	?	in NC
13	✓		✓		?	in NC
134	✓		✓	✓	?	in NC
12	✓	✓			in NC	in NC
124	✓	✓		✓	in NC	in NC
123	✓	✓	✓		in NC	in NC
1234	✓	✓	✓	✓	$\mathcal{O}(1)$	$\mathcal{O}(1)$

To exhibit the non-totalistic but rotation invariant rules, it is enough to differentiate the output when there are exactly two states at value 1:

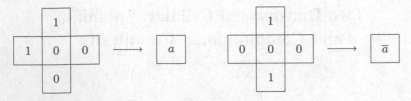

Fig. 1. Non-totalistic update for von Neumann neighborhood. $a \in \{0, 1\}$

For the automaton's dynamics we will consider two update models, synchronous (every site is updated at the same time) and asynchronous: sites are updated one by one in a prescribed order or equivalently following a permutation of the set of sites. In this case, we call the permutation a sequential updating scheme. From that we define the following decision problems:

Eventual-Prediction (EventPred)
Input: A finite configuration x of dimensions $n \times n$ and a site $u \in [n] \times [n]$ such that
$x_u = 0$.
Question: Does there exist $t^* > 0$ such that $(F^{t^*}(x))_u = 1$?

and

Asynchronous-Prediction (AsyncPred)
Input: A finite configuration x of dimensions $n \times n$, a site $u \in [n] \times [n]$ such that
$x_u = 0$.
Question: Does there exist a sequential updating scheme σ and $t^* > 0$ such that
$(F^{\sigma t^*}(x))_u = 1$.

For some of those problems we will exhibit their computational complexity as well as the tools developed to prove it. Further, for AsyncPred, we proved that the Strict Majority Automata is in *NC*. given an answer to the conjecture proposed by C. Moore.

References

1. Goles, E., Montealegre, P.: The complexity of the asynchronous prediction for the strict majority automata (2016, preprint)
2. Goles, E., Maldonado, D., Montealegre, P., Ollinger, N.: Two state totalistic freezing cellular automata and their complexity. In: Cellular Automata (2016). Exploratory paper
3. Moore, C.: Majority vote cellular automata, Ising dynamics, and P-completeness. J. Stat. Phys. **88**(3), 795–805 (1977)

Fixed Points in Boolean Networks

Adrien Richard

CNRS & Université de Nice Sophia Antipolis, Laboratoire I3S,
UMR CNRS 7271, 06900 Sophia-Antipolis, France
richard@unice.fr

A Boolean network is defined from a finite digraph G by associating to each vertex a binary variable and a local transition function, which depends on in-neighbors' variables. Dynamics are then obtained by applying the local transition functions, synchronously or asynchronously.

Boolean networks have many applications. For instance, they are classical models for gene networks. In this context, the interaction graph G is often known, or at least well approximated, while the actual dynamics are not. A natural question is then the following: *what can be said on the dynamics according to the interaction graph G only?*

In this presentation, we give partial answers, focusing on the maximum number of fixed points and some particular classes of networks, such as monotone networks.

References

1. Aracena, J., Richard, A., Salinas, L.: Number of fixed points and disjoint cycles in monotone boolean networks. SIAM J. Discrete Math. (2017, to appear)
2. Gadouleau, M., Richard, A., Riis, S.: Fixed points of boolean networks, guessing graphs, and coding theory. SIAM J. Discrete Math. **29**(4), 2312–2335 (2015)

Contents

Invited Papers

Strict Asymptotic Nilpotency
in Cellular Automata

Ville Salo$^{(\boxtimes)}$

University of Turku, Turku, Finland
vosalo@utu.fi

Abstract. We discuss the problem of which subshifts support strictly asymptotically nilpotent CA, that is, asymptotically nilpotent CA which are not nilpotent. The author talked about this problem in AUTOMATA and JAC 2012, and this paper discusses the (lack of) progress since. While the problem was already solved in 2012 on a large class of multi-dimensional SFTs, the full solutions are not known for one-dimensional sofics, multidimensional SFTs, and full shifts on general groups. We believe all of these questions are interesting in their own way, and discuss them in some detail, along with some context.

1 Introduction

A cellular automaton is nilpotent if every configuration is eventually mapped to the all-zero configuration. This notion is best known as an undecidable property of one-dimensional CA, see [1,6]. Here, we discuss its dynamical aspects. We study the relation of nilpotency and its asymptotic version in the setting of cellular automata on subshifts. Our main point is to state the following questions.

Question 1. Which one-dimensional sofic shifts support cellular automata which are asymptotically nilpotent but not nilpotent?

Question 2. Which multidimensional SFTs support cellular automata that are asymptotically nilpotent but not nilpotent? Do any?

Question 3. Which full shifts on countable groups support cellular automata that are asymptotically nilpotent but not nilpotent? Do any?

Below, these are Questions 4, 6 and 9, respectively. They are discussed in their natural contexts, and we try to include some informal guesses about what the solutions might look like. We also ask other questions, and share some lemmas and examples.

2 Nilpotency on Multidimensional Full Shifts

We begin with an introduction of the problem and its full solution in the classical multidimensional full shift setting.

© IFIP International Federation for Information Processing 2017
Published by Springer International Publishing AG 2017. All Rights Reserved
A. Dennunzio et al. (Eds.): AUTOMATA 2017, LNCS 10248, pp. 3–15, 2017.
DOI: 10.1007/978-3-319-58631-1_1

Let $\Sigma \ni 0$ be a finite alphabet.[1] The *d-dimensional full shift* is the dynamical system $\Sigma^{\mathbb{Z}^d}$ where $\mathbb{Z}^d$ acts by translations $\sigma_v(x)_u = x_{u+v}$. A *cellular automaton* is a continuous shift-commuting function $f : \Sigma^{\mathbb{Z}^d} \to \Sigma^{\mathbb{Z}^d}$. Cellular automata can be characterized concretely as follows: if f is a cellular automaton, then there is a *local rule* $f_{\mathrm{loc}} : \Sigma^F \to \Sigma$ where $F \subset \mathbb{Z}^d$ is a finite *neighborhood* such that $f(x)_v = f_{\mathrm{loc}}(\sigma_v(x)|_F)$ for all $x \in \Sigma^{\mathbb{Z}^d}$ and $v \in \mathbb{Z}^d$.

A cellular automaton is *nilpotent* if it is a root of the trivial (constant-zero) cellular automaton, that is, there exists $n \in \mathbb{N}$ such that $f^n(x) = 0^{\mathbb{Z}}$ for all $x \in \Sigma^{\mathbb{Z}^d}$.

The second kind of nilpotency we are interested in is *asymptotic nilpotency*. A cellular automaton is asymptotically nilpotent if every configuration converges to the same point in the limit. More precisely,

$$\forall x \in \Sigma^{\mathbb{Z}^d} : \forall v \in \mathbb{Z}^d : \exists n_0 \in \mathbb{N} : \forall n \geq n_0 : f^n(x)_v = 0.$$

Clearly a nilpotent cellular automaton is asymptotically nilpotent. If a cellular automaton is asymptotically nilpotent but not nilpotent, then it is *strictly asymptotically nilpotent* or *SAN*. This cannot happen on a d-dimensional full shift:

Theorem 1. *A cellular automaton $f : \Sigma^{\mathbb{Z}^d} \to \Sigma^{\mathbb{Z}^d}$ is asymptotically nilpotent if and only if it is nilpotent.*

This is proved for $d = 1$ in [5], and for general d in [12]. One might think the proof is direct compactness argument, but in fact the first relies on the geometry of $\mathbb{Z}$ and the second on algebraic properties of $\mathbb{Z}^d$. We discuss the ideas behind these proofs in Sect. 6.

3   Nilpotency from a Subset of Configurations

In this section, we show examples of how taking our initial configurations from a noncompact set can lead to SAN-like phenomena. These will be used as the basis of constructions in compact settings.

Let $f : \Sigma^{\mathbb{Z}^d} \to \Sigma^{\mathbb{Z}^d}$ be a cellular automaton. Let $X \subset \Sigma^{\mathbb{Z}}$ be a subset of $\Sigma^{\mathbb{Z}}$. We say f is *weakly nilpotent on X* if and only if

$$\forall x \in X : \exists n : f^n(x) = 0,$$

nilpotent on X if and only if $f^n(X) = \{0^{\mathbb{Z}^d}\}$ for some $n \in \mathbb{N}$, and *asymptotically nilpotent on X* if $f^n(x) \longrightarrow 0^{\mathbb{Z}^d}$ for all $x \in X$.

If $X = \Sigma^{\mathbb{Z}^d}$, then clearly f is nilpotent (in the sense of the previous section) if and only if it is nilpotent on X (in the sense of this section). It is also known that f is weakly nilpotent on $\Sigma^{\mathbb{Z}^d}$ if and only if it is nilpotent on $\Sigma^{\mathbb{Z}^d}$.

We note two trivial examples that are useful to keep in mind:

---

[1] Nilpotency discussions are at their clearest when subshifts are *pointed*, that is, they have a special point $0^{\mathbb{Z}^d}$ as part of their structure, and all nilpotency is toward this special point.

Example 1 (The shift). Let $X \subset \Sigma^{\mathbb{Z}^d}$ contain only *finite* configurations, that is, only configurations $x \in \Sigma^{\mathbb{Z}^d}$ such that the *support* $\mathrm{supp}(x) = \{v \in \mathbb{Z}^d \mid x_v \neq 0\}$ is finite. Then the shift map σ_v (for any nonzero vector v) is asymptotically nilpotent on X, but is weakly nilpotent on X if and only if $X = \{0^{\mathbb{Z}^d}\}$.

Example 2 (The spreading state CA). Let $S \subset \mathbb{Z}^d$ be any generating set for the group $\mathbb{Z}^d$ such that S contains the all-zero vector $0^d \in \mathbb{Z}^d$. Consider the CA f with neighborhood S and local rule $f_{\mathrm{loc}}(P) = a$ where $a = 0$ if $P_v = 0$ for some $v \in S$ and $a = P_{0^d}$ otherwise. Thus, 0 is a *spreading state* that spreads into every cell that 'sees it'. If a configuration x contains 0, then f is asymptotically nilpotent on $\{x\}$, and it is nilpotent on $\{x\}$ if and only if 0 occurs in a *syndetic set*, that is, in every translate of a ball of large enough radius. Now the following are easy to see:

- f is weakly nilpotent but not nilpotent on the (dense) set of all finite points,
- f is asymptotically nilpotent on X when X is the (dense) set of generic configurations for some full support ergodic measure μ, but is not weakly nilpotent on this set for any such μ.

The second item is based on the fact that in a generic point for an ergodic measure, we (by definition) see every pattern with the correct frequency, so in particular we see every pattern.

For finite configurations, these trivial examples are sufficient for our purposes, but we note that erasing/eroding finite patterns is a much-studied topic in cellular automata, and there are several interesting constructions and results in this setting. Perhaps the most famous eroder is the GKL automaton [4].

Periodic points are another important non-compact set of starting configurations.

Theorem 2. *There is a CA on a one-dimensional full shift which is nilpotent on periodic configurations but is not nilpotent.*

The existence of such CA is a direct corollary of the undecidability of nilpotency [1,6], as nilpotency is semi-decidable, and non-nilpotency on periodic configurations is semi-decidable. There are also very simple examples if we consider only periodic configurations whose periods are restricted, and they are actually enough for our application in Sect. 6: the XOR CA $f : \mathbb{Z}_2^{\mathbb{Z}} \to \mathbb{Z}_2^{\mathbb{Z}}$ defined by $f(x)_i = x_i + x_{i+1}$ (where $\mathbb{Z}_2 = \mathbb{Z}/2\mathbb{Z}$) is well-known to be nilpotent on periodic configurations with period of the form 2^n.

4   Nilpotency on Multidimensional Subshifts

In the previous section we considered nilpotency when starting from a noncompact set of configurations. Perhaps more natural is to consider the initial set of configurations X to be a *subshift*, that is, $\sigma_v(X) = X$ for all $v \in \mathbb{Z}^d$ and X is topologically closed, or equivalently X is defined by a set of forbidden patterns.

In this section, we concentrate on the case when X is a subshift and is also closed under f, so that $f : X \to X$ is the restriction of a cellular automaton on $\Sigma^{\mathbb{Z}^d}$; this is the setting of cellular automata on multidimensional subshifts, and such f are precisely the continuous shift-commuting functions on X.

4.1 SFTs

First, consider the case $d = 1$. Let $X \subset \Sigma^{\mathbb{Z}}$ be a subshift of finite type or *SFT*, that is, a closed shift-invariant subset of $\Sigma^{\mathbb{Z}}$ obtained by forbidding a finite set of words. Such X is *conjugate*[2] to the set of paths in a finite graph [8]. It is shown in [5,12] that in this setting there are no SAN cellular automata.

Theorem 3 (Corollary 1 in [12]). *Let $X \subset \Sigma^{\mathbb{Z}}$ be an SFT. Then a cellular automaton f on X is nilpotent if and only if it is asymptotically nilpotent.*

One can try to generalize this to SFTs in higher dimensions. SFTs of $\Sigma^{\mathbb{Z}^d}$ are defined like in one dimension, by forbidding finitely many patterns from occurring (and such systems are conjugate to tiling systems induced by finitely many tiles).

Theorem 4 (Theorem 4 in [12]). *Let $X \subset \Sigma^{\mathbb{Z}^d}$ be an SFT where finite points are dense. Then a cellular automaton f on X is nilpotent if and only if it is asymptotically nilpotent.*

The denseness of finite points is not a property that is often assumed from multidimensional SFTs. A more commonly used gluing property is so-called strong irreducibility. Many other gluing properties have been defined, and some are listed for example in [2].

Definition 1. *Let $X \subset \Sigma^{\mathbb{Z}^d}$ be a subshift. We say X is strongly irreducible if there exists $m \in \mathbb{N}$ such that for any $y, z \in X$ and any finite sets $N, N' \subset \mathbb{Z}^d$ with $\min\{|v - v'| \mid v \in N, v' \in N'\} \geq m$, there exists a point $x \in X$ with $x_N = y_N$ and $x_{N'} = z_{N'}$.*

By compactness, the sets N and N' can then be taken infinite as well, and we easily obtain the following lemma.

Lemma 1. *Let $X \subset \Sigma^{\mathbb{Z}^d}$ be a strongly irreducible SFT and let $x \in X$. Then the points asymptotic to x are dense in X.*

Since a subshift has to have the point $0^{\mathbb{Z}^d}$ in order to support asymptotically nilpotent cellular automata, we see that in our case of interest, strong irreducibility is a stronger requirement than the density of 0-finite points:

Corollary 1. *Let $X \subset \Sigma^{\mathbb{Z}^d}$ be a strongly irreducible SFT. Then a cellular automaton f on X is nilpotent if and only if it is asymptotically nilpotent.*

---

[2] Equal up to shift-commuting homeomorphism.

I do not know if any such property is required.

Question 4. Is there an SFT $X \subset \Sigma^{\mathbb{Z}^d}$ for some $d \geq 2$ and a CA $f : X \to X$ such that f is asymptotically nilpotent but not nilpotent?

If the answer is yes, then the solution would presumably involve constructing an SFT where every configurations sets up some zones where we eventually have only zeroes, in some controlled way. There are many known constructions that might allow this [3,15], but since the SFT must be closed under f, this seems difficult.

It is known that the shift is not SAN on any $\mathbb{Z}^2$-SFT [10,13]. In [13], this is shown using topological arguments: by passing to a minimal subsystem (in a technical sense) we can always find certain abstract 'spaceships', which prevent nilpotency. The proof is very specific to $d = 2$, and does not seem to extend to $d = 3$.

In fact, a cellular automaton on an SFT can be seen as a shift on a higher-dimensional SFT by looking at its spacetime diagrams, and thus a positive answer to Question 4 would imply a positive answer to the following (where by the previous paragraph we can just as well restrict to $d \geq 3$):

Question 5. Is there an SFT $X \subset \Sigma^{\mathbb{Z}^d}$ for some $d \geq 3$ such that X contains at least two points, and σ_v is asymptotically nilpotent for some $v \in \mathbb{Z}^d$?

A shift map is nilpotent on an SFT X (as a cellular automaton) if and only if $X = \{0^{\mathbb{Z}^d}\}$, so we may replace the assumption that σ_v is not nilpotent by the assumption that X has at least two points. The direction v of the shift does not matter, since SFTs are closed under rotating the lattice by elements of $GL(n, \mathbb{Z})$.

4.2 Sofics (and Beyond)

The proof of Theorem 4 in [12] in fact does not require that X is an SFT, but only the property that taking the union of the supports of two configurations of X gives a point in X assuming the supports have large enough distance. This is called *zero-gluing* in [13], where nilpotency is studied in the nondeterministic setting.

SFTs are always zero-gluing, but *sofic shifts*, which are images of SFTs under cellular automata (which clearly generalizes SFTs), are not. Let us show how the trivial idea of shifting finite points from the last section leads to an equally trivial example of a sofic shift where we can have SAN maps. We remark that sofic shifts on $\mathbb{Z}$ are precisely the subshifts whose language is regular, and precisely the subshifts that are defined by a regular language of forbidden words.

Example 3 (One-one). Let $\Sigma = \{0,1\}$ and let $X \subset \Sigma^{\mathbb{Z}}$ be defined by the forbidden patterns 10*1. This is the *one-one* subshift. Then X is a countable sofic shift. Then σ is asymptotically nilpotent but not nilpotent on X: on the configuration x^i where $x_j^i = 1 \iff i = j$, the cell at the origin contains 0 after i steps if $i \geq 0$, so σ is not nilpotent. However, there can only be one occurrence of 1 in a configuration of X, so σ is asymptotically nilpotent.

If X is sofic and has points with infinite support, then the shift is not SAN (because it is not even asymptotically nilpotent). Using a different CA, there is an example that is *topologically transitive*, that is for any words u, v that occur in points of X, there exists a word w such that uwv occurs in a point of X.

Example 4. Let $\Sigma = \{\leftarrow, \rightarrow, 0\}$ and let X be the sofic shift with forbidden patterns the regular language $\leftarrow 0^* \leftarrow + \rightarrow 0^* \rightarrow$. Let f be the cellular automaton that moves every $\leftarrow$ to the left and every $\rightarrow$ to the right, removing both on collision. Then f is asymptotically nilpotent but not nilpotent.

Of course, similar examples are obtained on $\mathbb{Z}^d$ – the one-one example directly generalizes to higher dimensions, and any one-dimensional subshift can be turned into a higher-dimensional one by having an independent copy of it in every $\mathbb{Z}$-coset.

There are also sofic shifts which are not zero-gluing (and thus are in particular not SFTs), but which do not support SAN maps. The following result can be shown with the proof of [5].

Example 5 (The even and odd shifts). Let $\Sigma = \{0, 1\}$ and let $X \subset \Sigma^{\mathbb{Z}}$ be the subshift with forbidden patterns the regular language $1(00)^*1$, and $Y \subset \Sigma^{\mathbb{Z}}$ the one with forbidden patterns $10(00)^*1$. The subshift X is called the *odd shift*, and Y the *even shift*. Both are proper sofic, and neither supports SAN CA.

Intuitively, what is going on is that the even and odd shifts are 'almost' zero-gluing, in that two finite configurations can be glued together, up to shifting one of them by one, allowing the same constructions as are used in [5].

Every known one-dimensional sofic shift that does allow SAN maps is based on either the one-one or the idea of colliding particles, and subshifts that are, intuitively, 'almost' of finite type do not support SAN CA. Can one find a characterization of sofic shifts allowing such behavior?

Question 6. Let $X \subset \Sigma^{\mathbb{Z}}$ be a sofic shift. Under what conditions does there exist a cellular automaton on X which is asymptotically nilpotent but not nilpotent?

Question 7. Let $X \subset \Sigma^{\mathbb{Z}}$ be a subshift. Are there natural specification, mixing, or gluing properties that forbid the existence of SAN maps X? What properties of X are needed for the proofs of [5,12] to go through?

Another classification question is which sofic shifts have an undecidable nilpotency problem. In [14] it is shown that nilpotency is decidable on countable sofic shifts, and we know it is undecidable on full shifts, so this class is something in-between.

5   Nilpotency as Uniform Convergence to a Point

We now set up a more general framework for nilpotency, so that we do not need to give new definitions every time we generalize our model, and so that we can discuss nilpotency also in positive-dimensional settings.

By definition, a cellular automaton is asymptotically nilpotent if and only if the limit of every point in the action of the CA is the all-zero point. We can characterize nilpotency in terms of this convergence:

Lemma 2. *Let $X \subset \Sigma^{\mathbb{Z}^d}$ be any subshift with $0^{\mathbb{Z}^d} \in X$ and let $f : X \to X$ be a CA. Then f is nilpotent if and only if $f^n(x) \longrightarrow 0^{\mathbb{Z}^d}$ uniformly in $x \in X$.*

Inspired by this, we give the following definitions: If (X, f) is an $\mathbb{N}$-dynamical system (X is a topological space and $f : X \to X$ is a continuous function), then we say it is *nilpotent*, if there is a point $0 \in X$ such that for some $n \in \mathbb{N}$, $f^n(x) = 0$ for all $x \in X$, *asymptotically nilpotent* or *AN* if there is a point $0 \in X$ such that $f^n(x) \longrightarrow 0$ for all $x \in X$, and *uniformly asymptotically nilpotent* or *UAN* if this convergence is uniform over X.

We have

$$\text{nilpotent} \implies \text{UAN} \implies \text{AN},$$

and we will see these implications are strict in general.

A system is *non-uniformly asymptotically nilpotent* or *NUAN* if it is asymptotically nilpotent but not uniformly so, and we define *SAN* as before as being asymptotically nilpotent but not nilpotent:

$$\text{SAN} = \text{AN} \wedge \neg\text{nilpotent}, \qquad \text{NUAN} = \text{AN} \wedge \neg\text{UAN}$$

There are many systems that are uniformly asymptotically nilpotent but not nilpotent, even in the zero-dimensional setting, but the previous lemma shows a CA is nilpotent if and only if it is UAN, so a cellular automaton is SAN if and only if it is NUAN.[3] In general, for a dynamical system we can only say

$$\text{NUAN} \implies \text{SAN}.$$

The questions we are interested in are of the following type: Given a class D of dynamical systems, are there NUAN/SAN systems in D? In this article, we mostly study the case where we fix a dynamical system (a subshift) and let D be its endomorphisms, but we can consider the NUAN/SAN behavior more generally.

We make some basic remarks (see also [5]) and give some examples: A $\mathbb{Z}$-subshift X, seen as an $\mathbb{N}$-system with the action of σ is, by definition, AN if and only if the cellular automaton σ is asymptotically nilpotent on it, and this happens precisely when all points of X contain only zeros in their eventual right tail. Such a subshift is only UAN if and only it is nilpotent if and only if $X = \{0^{\mathbb{Z}}\}$. Thus the notions NUAN and SAN agree for $\mathbb{Z}$-subshifts as well.

For an $\mathbb{N}$-subshift X, the characterization of AN is the same. Such X is UAN if and only if it is nilpotent if and only if there is a bound $m \in \mathbb{N}$ such that $x \in X \implies \forall n \geq m : x_n = 0$. These are precisely the $\mathbb{N}$-subshifts which are finite and contain only finite points. Thus the notions NUAN and SAN agree for $\mathbb{N}$-subshifts.

---

[3] See, however, Example 6.

If $X \subset \Sigma^{\mathbb{Z}^d}$ is a subshift, then a cellular automaton $f : X \to X$ is AN if and only if its *trace* $\{y \in \Sigma^{\mathbb{N}} \mid \exists x \in X : y_n = f^n(x)_{0^d}\}$ is AN as a one-dimensional subshift, and f is nilpotent if and only if its trace is.

As we have seen, there are subshifts which are SAN (equivalently NUAN), for example the one-one subshift. The one-sided one-one subshift (the set of right tails of the one from the previous section) can also be described as follows: Let $\dot{\mathbb{N}} = \mathbb{N} \cup \{\infty\}$ be the one-point compactification of $\mathbb{N}$, with its usual topology. Then the map f defined by $f(n) \mapsto n - 1$ (with $\infty - 1 = \infty$) for $n \geq 1$ and $f(0) = \infty$ is continuous on $\dot{\mathbb{N}}$, and $(\dot{\mathbb{N}}, f)$ is NUAN.

The two-sided one-one subshift can similarly be seen as the subtraction homeomorphism on $\dot{\mathbb{Z}}$. There is also a more geometric way to implement this idea: Seeing the circle S^1 as $[0, 1]$ with 0 and 1 identified, the map $f(x) = x^2$ is well-defined and continuous. In this system (S^1, f), every point converges to 0, but this convergence cannot be uniform, as f is a homeomorphism (so every point has a preimage). This example is also NUAN.

The map $x \mapsto x/2$ also gives us continuous dynamics on $[0, 1]$. This map is not nilpotent, but is UAN. Thus, it is SAN but not NUAN.

6   Cellular Automata on Graphs and Groups

Next, we generalize in another direction, and replace $\mathbb{Z}^d$ by a graph.

Let G be any *road-colored graph* whose edges are colored with colors from a finite set S in such a way that for every vertex $v \in G$, for each $s \in S$ there is a unique edge with v as initial vertex and s as label. The terminal vertex of this unique edge is denoted by vs. If $u \in S^*$, we write $vu = (\cdots((vu_0)u_1) \cdots u_{|u|-1})$ (with $v\epsilon = v$ where ϵ is the empty word).

Write Σ^G for the set of colorings of the nodes of G. A *cellular automaton* on Σ^G is a function $f : \Sigma^G \to \Sigma^G$ such that for some function $f_{\text{loc}} : \Sigma^F \to \Sigma$ where $F \subset S^*$ is finite, we have $f(x)_v = f_{\text{loc}}(a \mapsto x_{va})$ for all $x \in \Sigma^G$ and $a \in F$.

An important example are cellular automata on groups: Let G be a group generated by a finite set S, and Σ a finite alphabet. Then G acts on Σ^G by left translations $g \cdot x_h = x_{g^{-1}h}$, and continuous functions $f : \Sigma^G \to \Sigma^G$ commuting with them are precisely the functions that are cellular automata in the sense of the previous definition on any Cayley graph of G, that is, there always exist $f_{\text{loc}} : \Sigma^F \to \Sigma$ such that

$$f(x)_v = f_{\text{loc}}(x_{vs_1}, x_{vs_2}, \ldots, x_{vs_k})$$

where $s_1, s_2, \ldots, s_k \in G$. Setting $G = \mathbb{Z}^d$, we obtain the classical setting of multidimensional CA. For cellular automata on groups, UAN is equivalent to nilpotency. Thus, in this setting SAN and NUAN are equivalent concepts as well.

One can similarly define cellular automata on monoids: If M is a monoid generated by a set S, then M has right Cayley graph with edges (m, ms) where $s \in S$, and we can consider cellular automata on this graph. Setting $M = \mathbb{N}$, this corresponds to the usual one-sided cellular automata.

Example 6. Let $f : \Sigma^{\mathbb{Z}} \to \Sigma^{\mathbb{Z}}$ be a cellular automaton which is asymptotically nilpotent on periodic points, but not nilpotent, see Theorem 2. Let C_n be the graph with n nodes in a single cycle, with label 1 on each edge. Let $G = \bigcup_n C_n$ be a disjoint union of these graphs, and let $g : \Sigma^G \to \Sigma^G$ be the cellular automaton with the same local rule as f. Then clearly g is asymptotically nilpotent. In fact, g is UAN: for every cycle, there is a bounded time after which it contains only zeroes. Nevertheless, g is not nilpotent, as f is not. It follows that this example is not nilpotent, but is UAN. Thus, it is SAN but not NUAN.

Intuitively, this example comes from the fact that the graph is not homogenous, in the sense that the neighborhoods of different nodes look different. If we could 'translate the graph' and take limits of this translation process, we would in a natural sense obtain copies of $\mathbb{Z}$ where f is not even asymptotically nilpotent. We do not formalize this idea here.

Example 6 can be seen as a cellular automaton on a group action, as the cellular automaton we define commutes with the action of $\mathbb{Z}$ that rotates the information on each C_n. We do not formalize this idea either.

While we can have SAN in the setting of CA on graphs (by the previous example), I do not know whether NUAN behavior is possible.

Question 8. Is there a road-colored graph G and an alphabet Σ such that some CA $f : \Sigma^G \to \Sigma^G$ is NUAN?

The main question is whether we can obtain SAN maps in the group setting:

Question 9. Is there a countable group G and an alphabet Σ such that some CA $f : \Sigma^G \to \Sigma^G$ is SAN?

6.1   What Works on General Groups?

Many of the arguments of [5,12] work on every group, but have only been stated in the $\mathbb{Z}^d$ case. The first is from [5], and states that there are some patterns that empty a particular cell (or set of cells) forever, in any context.

Lemma 3. *Suppose G is a countable group and $f : \Sigma^G \to \Sigma^G$ is an asymptotically nilpotent CA. Then for every neighborhood V of 0^G there exists a nonempty open set U and $n_0 \in \mathbb{N}$ such that $x \in U \implies \forall n \geq n_0 : f^n(x) \in V$.*

The second and third lemmas deal with mortal finite configurations. If $f : X \to X$ is a CA and $x \in X$, then x is *mortal* if f is nilpotent on $\{x\}$, that is, $f^n(x) = 0^G$ for some $n \in \mathbb{N}$. The proof of both lemmas below are based on the observation that in an asymptotically nilpotent CA, every cell (or set of cells) must regularly visit the all-zero cylinder.

The second lemma states that if a finite configuration does not spread under an asymptotically nilpotent CA, then it is mortal.

Lemma 4. *Suppose G is a countable group and $f : \Sigma^G \to \Sigma^G$ is an asymptotically nilpotent CA. If x is a finite configuration such that the set $\{v \in G \mid \exists n : f^n(x)_v \neq 0\}$ is finite, then x is mortal.*

The following is proved like Lemma 2 of [12].

Lemma 5. *Suppose G is a countable group, $f : \Sigma^G \to \Sigma^G$ is an asymptotically nilpotent CA and there is a dense set of finite mortal configurations. Then f is nilpotent.*

We give an intuitive outline of the proof of the case $G = \mathbb{Z}$ of Theorem 1 in [5], highlighting how the geometry of $\mathbb{Z}$ is used: Suppose we have an asymptotically nilpotent CA. We use Lemma 3 to get a word w which blocks information flow through its central coordinate. *One half of such word (after a few iterations) must then block information flow from one side.* Sticking such halves around a finite configuration, the configuration becomes stuck in a finite segment, and it must then be mortal by Lemma 4. This means there is a dense set of mortal finite configurations, and we conclude with Lemma 5.[4]

In this proof, we make essential use of the fact $\mathbb{Z}$ has multiple ends: Cutting w in half effectively splits the group in two, and the CA will never see over the gap. However, everything else follows from general arguments.

In the $d \geq 2$ case, the same idea does not work directly, as the blocking words w are replaced by blocking patterns, and they need not block information flow in any essential way. The new idea in [12] is that periodizing a point makes the cellular automaton simulate a one-dimensional CA, and we already know the result for such CA. The periodization must be doable in any direction, so we make essential use of the fact that there are quotient maps $\mathbb{Z}^d \to \mathbb{Z}$ 'for every dimension d', which is almost the definition of $\mathbb{Z}^d$. In this sense, we make essential use of the algebraic structure of $\mathbb{Z}^d$.

For some specific groups, it is plausible that SAN behavior can be ruled out easily, by a more assiduous application of these ideas. In the case of free groups (which also have multiple ends), one can directly attempt to mimic the proof of [5], while in the case of the Heisenberg group, one can try to mimic the periodization idea of [12].

Conjecture 1. If G is a free group or G is the Heisenberg group, then there are no SAN cellular automata on any full shift on G.

7 CA with Very Sparse Spacetime Diagrams

In this section, we list some results of [15]. The construction of [15] gives insight into why it is difficult to show that SAN behavior is impossible: we can 'almost' have it on a full shift, in the sense that a non-nilpotent cellular automaton can have both very sparse rows and very sparse columns in the limit.

Consider the full shift $\Sigma^{\mathbb{Z}}$. Then the *Besicovitch pseudometric* is defined by

$$d_B(x, y) = \limsup_{n \to \infty} \frac{1}{2n + 1} |\{-n \leq i \leq n \mid x_i \neq y_i\}|.$$

---

[4] As we only want to emphasize what property of $\mathbb{Z}$ is used, we are of course omitting some technical details. Interested readers will find the details in the references.

Then $(\Sigma^{\mathbb{Z}}, d_B)$ is a topological (non-Hausdorff non-compact) space called the *Besicovitch space*, and if $f : \Sigma^{\mathbb{Z}} \to \Sigma^{\mathbb{Z}}$ is a cellular automaton on $\Sigma^{\mathbb{Z}}$ (in the usual sense), then f is also an endomorphism (continuous shift-commuting self-map) of $(\Sigma^{\mathbb{Z}}, d_B)$. We write d_C for the usual metric inducing the product topology on $\Sigma^{\mathbb{Z}}$.

We cite some results of [15] about nilpotency on the Besicovitch space.

Proposition 1. *Let $f : \Sigma^{\mathbb{Z}} \to \Sigma^{\mathbb{Z}}$ be a cellular automaton. Then*

– *f is nilpotent on $(\Sigma^{\mathbb{Z}}, d_B)$ if and only if it is nilpotent on $(\Sigma^{\mathbb{Z}}, d_C)$, and*
– *if f is AN on $(\Sigma^{\mathbb{Z}}, d_B)$, then it is UAN on $(\Sigma^{\mathbb{Z}}, d_B)$.*

These follow from [15, Proposition 41] and [15, Lemma 43], respectively. The second item shows that there are no cellular automata that are NUAN over the Besicovitch space.

The main result of [15] is the construction of a CA with very sparse spacetime diagrams, and the following is one of the corollaries of the construction:

Theorem 5. *There is a cellular automaton which is UAN on $(\Sigma^{\mathbb{Z}}, d_B)$ but is not nilpotent on $(\Sigma^{\mathbb{Z}}, d_C)$.*

Thus, there are cellular automata that are SAN over the Besicovitch space.

Asymptotic nilpotency for the Besicovitch metric is 'spatial nilpotency', that is, most cells being zero on every row. Thus, to formulate this notion, we need a notion of 'space'. For the temporal direction, we can formulate a notion of nilpotency in density directly: If $f : X \to X$ is a dynamical system and $0 \in X$, then f is *asymptotically nilpotent in temporal density* or $ANTD^5$ if

$$\forall \epsilon > 0 : \forall x \in X : \liminf_{n \to \infty} \frac{1}{n} |\{0 \le i < n \mid d(f^i(x), 0) < \epsilon\}| = 1.$$

If the convergence is uniform in X for all ϵ, we use the obvious acronym *UANTD*.

Proposition 2. *Let $f : \Sigma^{\mathbb{Z}} \to \Sigma^{\mathbb{Z}}$ be a cellular automaton. Then*

– *f has a unique invariant measure if and only if it is ANTD, and*
– *if f is ANTD then it is UANTD.*

The first observation is well-known, see [15, Proposition 5] for a proof. It is the reason why ANTD cellular automata are usually called *uniquely ergodic*, since this is the term for having a unique invariant measure in ergodic theory. The second condition says that there are no CA that are 'non-uniformly asymptotically nilpotent in temporal density', and it is also shown in [15, Proposition 5].

The construction of [15] also gives a CA where the temporal density of zeroes is high:

Theorem 6. *There exists a cellular automaton $f : \Sigma^{\mathbb{Z}} \to \Sigma^{\mathbb{Z}}$ which is ANTD but not nilpotent.*

---

[5] It would be nice to call this just 'asymptotic nilpotency in density', but this is used in [15] in the spatial sense.

Thus, there are cellular automata that are 'strictly asymptotically nilpotent in temporal density'.

We can formalize the property of [15] that the density of zeroes becomes high from every starting configuration also using measures: the CA of [15] proving Theorem 5 also has the property that every shift-invariant measure converges to the Dirac measure at zero in the iteration of the CA. In other words, there is a non-nilpotent CA $f : \Sigma^{\mathbb{Z}} \to \Sigma^{\mathbb{Z}}$ which, acting over the space of shift-invariant probability measures on $\Sigma^{\mathbb{Z}}$, is asymptotically nilpotent toward the Dirac measure at $0^{\mathbb{Z}}$.

We mention that for cellular automata, there is also another well-known kind of nilpotency on the space of measures, namely randomization:

Example 7. Let $f : \mathbb{Z}_2^{\mathbb{Z}} \to \mathbb{Z}_2^{\mathbb{Z}}$ be the two-neighbor XOR $f(x)_i = x_i + x_{i+1}$. Then f acts on the space of shift-invariant measures $\mathcal{M}_\sigma$ on $\mathbb{Z}_2^{\mathbb{Z}}$. It is known that f is *randomizing in density* [7,9,11] in the sense that $f^n(\mu)$ converges weakly to the uniform Bernoulli measure 'except for a set of times of density zero', whenever μ is a full-support Bernoulli measure.[6] Combining the terminology of Sect. 3 with that of this section means precisely that f is ANTD on the set of full-support Bernoulli measures, with nilpotency towards the uniform measure.

Acknowledgements. We thank Pierre Guillon and Ilkka Törmä for their comments on the draft.

References

1. Aanderaa, S.O., Lewis, H.R.: Linear sampling, the ∀∃∀ case of the decision problem. J. Symbolic Logic **39**, 519–548 (1974)
2. Boyle, M., Pavlov, R., Schraudner, M.: Multidimensional sofic shifts without separation and their factors. Trans. Am. Math. Soc. **362**(9), 4617–4653 (2010)
3. Durand, B., Romashchenko, A., Shen, A.: Fixed-point tile sets and their applications. J. Comput. Syst. Sci. **78**(3), 731–764 (2012)
4. Gács, P., Kurdyumov, G.L., Levin, L.A.: One-dimensional uniform arrays that wash out finite islands. Problemy Peredachi Informatsii **14**(3), 92–96 (1978)
5. Guillon, P., Richard, G.: Asymptotic behavior of dynamical systems and cellular automata. ArXiv e-prints, April 2010
6. Kari, J.: The nilpotency problem of one-dimensional cellular automata. SIAM J. Comput. **21**(3), 571–586 (1992)
7. Lind, D.A.: Applications of ergodic theory and sofic systems to cellular automata. Physica D **10**(1–2), 36–44 (1984)
8. Lind, D., Marcus, B.: An Introduction to Symbolic Dynamics and Coding. Cambridge University Press, Cambridge (1995)
9. Miyamoto, M.: An equilibrium state for a one-dimensional life game. J. Math. Kyoto Univ. **19**(3), 525–540 (1979)
10. Pavlov, R., Schraudner, M.: Classification of sofic projective subdynamics of multidimensional shifts of finite type. Trans. Am. Math. Soc. **367**(5), 3371–3421 (2015)

---

[6] The initial set of measures can be extended considerably, but we do not know a published reference dealing with a class that is obviously closed under f.

11. Pivato, M.: The ergodic theory of cellular automata. Int. J. Gen. Syst. **41**(6), 583–594 (2012)
12. Salo, V.: On Nilpotency and Asymptotic Nilpotency of Cellular Automata. ArXiv e-prints, May 2012
13. Salo, V.: Subshifts with sparse projective subdynamics. ArXiv e-prints, May 2016
14. Salo, V., Törmä, I.: Computational aspects of cellular automata on countable sofic shifts. In: Mathematical Foundations of Computer Science 2012, pp. 777–788 (2012)
15. Törmä, I.: A uniquely ergodic cellular automaton. J. Comput. Syst. Sci. **81**(2), 415–442 (2015)

Regular Papers

Infinite Two-Dimensional Strong Prefix Codes: Characterization and Properties

Marcella Anselmo[1], Dora Giammarresi[2(✉)], and Maria Madonia[3]

[1] Dipartimento di Informatica, Università di Salerno,
Via Giovanni Paolo II, 132, 84084 Fisciano, SA, Italy
manselmo@unisa.it
[2] Dipartimento di Matematica, Università Roma "Tor Vergata",
via della Ricerca Scientifica, 00133 Roma, Italy
giammarr@mat.uniroma2.it
[3] Dipartimento di Matematica e Informatica,
Università di Catania, Viale Andrea Doria 6/a, 95125 Catania, Italy
madonia@dmi.unict.it

Abstract. A two-dimensional code is defined as a set of rectangular pictures over an alphabet Σ such that any picture over Σ is tilable in at most one way with pictures in X. It is in general undecidable whether a set of pictures is a code, even in the finite case. Recently, finite strong prefix codes were introduced in [3] as a family of decidable picture codes. In this paper we study infinite strong prefix codes and give a characterization for the maximal ones based on *iterated extensions*. Moreover, we prove some properties regarding the measure of these codes.

Keywords: Two-dimensional languages · Prefix codes · Measure

1 Introduction

Extending the theory of formal (string) languages to two dimensions is a very interesting and challenging task. Our motivations are mainly theoretical but, as formal language theory had very significant impact in several applications, we expect that results on two-dimensional languages will be exploited in practical fields like image processing, pattern recognition and matching.

A two dimensional word, or picture, is a rectangular array of symbols taken from a finite alphabet Σ; a two-dimensional language is thus a subset of Σ^{**}. The notion of finite state recognizability can be transferred into a two-dimensional (2D) world in different ways (e.g. [10,15,17,19,22–24]). A crucial difference with the string language theory is that in two dimensions many problems become undecidable and even for finite-state recognizability we loose the equivalence between determinism and non-determinism [2,6,17].

Partially supported by INdAM-GNCS Project 2017, FARB Project ORSA138754 of University of Salerno and FIR Project 375E90 of University of Catania.

© IFIP International Federation for Information Processing 2017
Published by Springer International Publishing AG 2017. All Rights Reserved
A. Dennunzio et al. (Eds.): AUTOMATA 2017, LNCS 10248, pp. 19–31, 2017.
DOI: 10.1007/978-3-319-58631-1_2

In the theoretical study of formal string languages, string codes have been always a relevant subject of research, also because of their applications to practical problems (see [14] for complete references). An important and easy-to-construct class of string codes are prefix codes. Recall that a set S of strings is called prefix if inside S no word is (left-)prefix of another one. It holds that any prefix set of words is also a code, referred to as a prefix code. The notion of code can be intuitively and naturally transposed to two-dimensional objects by exploiting the notion of unique tiling decomposition. Several attempts of developing a formal theory of *two-dimensional codes* have been done by using polyominoes (connected two-dimensional figures, not necessarily rectangular). Unfortunately, most of the published results show that in the 2D context we loose important properties. In [13] D. Beauquier and M. Nivat proved that the problem whether a finite set of polyominoes is a code is undecidable, and that the same result holds also for dominoes. Codes of other variants of polyominoes including bricks (i.e. labelled polyominoes) and pictures are also studied in [1,16,18,20,21] and further undecidability results are proved.

In [4,7], a new definition of picture code was introduced by referring to the operation of tiling star as defined in [24]; the tiling star of a set X is the set X^{**} of all pictures that are tilable (in the polyominoes style) by elements of X. Then, X is a code if any picture in X^{**} is tilable in a unique way. Unfortunately, it is again not decidable whether a finite language of pictures is a code. The aim was finding decidable subclasses of picture codes. For this, two definitions of prefix code of pictures have been proposed by associating to the pictures a preferred scanning direction from top-left corner towards the bottom-right one. Note that, moving to the 2D setting, the main concern is that if we delete a "prefix" from a picture (i.e. delete a rectangular portion starting at top-left corner) the remaining part is not in general a picture itself. As consequence, the proof techniques for string codes fail when transposed to two dimensions. Further generalizations to 2D of classes of string codes are presented in [8,11,12].

A first definition of two-dimensional prefix code is proposed in [4,7]. It is based on some special kind of polyominoes that have straight top border. A smaller class, referred to as the class of *strong prefix sets*, was then proposed in [3,9]; it is defined in a simpler way, it is easier to manage and more robust, while it preserves all positive features of the first definition. In order to prevent to start decoding a picture message in two different ways, no prefix-overlapping pictures are admitted in a strong prefix set. More precisely, any two pictures in the set cannot coincide in their common top-left part. Finite strong prefix sets are a decidable family of picture codes with a simple polynomial decoding algorithm. The results in [5,9], show a recursive procedure to construct all finite maximal strong prefix codes of pictures, starting from the "singleton" pictures containing only one alphabet symbol. The construction extends the literal representation of prefix codes of strings (cf. [14]). It is the starting point for most considerations in this paper.

All the mentioned results on two-dimensional codes regard *finite* codes, unless for some first examples of infinite codes of pictures in [8], in the framework of the deciphering delay. Here, the attention is devoted to the *infinite* strong prefix codes. We present a recursive definition of a family of languages based on the *iterated extensions*. We prove that all languages defined by iterated extensions are maximal strong prefix codes. Moreover, we show that, vice versa, any maximal strong prefix code can be obtained by iterated extensions. We investigate also the measure of such codes by associating a probability to each letter of the alphabet. We prove that, as in the string case, the measure of a two-dimensional strong prefix code is less than or equal to one. Nevertheless, we show that there exist infinite maximal strong prefix codes whose measure is strictly less than one and discuss the reason of this difference with the string case.

2 Preliminaries

We recall some definitions about two-dimensional languages (see [17]). A *picture* over a finite alphabet Σ is a two-dimensional rectangular array of elements of Σ. Given a picture p, $|p|_{row}$ and $|p|_{col}$ denote the number of rows and columns, respectively, while $size(p) = (|p|_{row}, |p|_{col})$ and $area(p) = |p|_{row} \times |p|_{col}$ denote the picture *size* and *area*, respectively. We also consider all the empty pictures that correspond to all pictures of size $(m, 0)$ or $(0, n)$. The set of all pictures over Σ of fixed size (m, n) is denoted by $\Sigma^{m,n}$. The set of all pictures over Σ is denoted by Σ^{**} while Σ^{++} refers to the set Σ^{**} without the empty pictures. A *two-dimensional language* (or *picture language*) over Σ is a subset of Σ^{**}. Any string on Σ can be viewed as a one-row picture in Σ^{**}. With a little abuse of notation, in the sequel, Σ will sometimes denote $\Sigma^{1,1}$, and a the corresponding picture in $\Sigma^{1,1}$.

In order to locate a position in a picture, it is necessary to put the picture in a reference system. The set of coordinates $dom(p) = \{1, 2, \ldots, |p|_{row}\} \times \{1, 2, \ldots, |p|_{col}\}$ is referred to as the *domain* of a picture p. We let $p(i, j)$ denote the symbol in p at coordinates (i, j). We assume the top-left corner of the picture to be at position $(1, 1)$, and fix the scanning direction for a picture from the top-left corner toward the bottom right one.

A *subdomain* of $dom(p)$ is a set d of the form $\{i, i + 1, \ldots, i'\} \times \{j, j + 1, \ldots, j'\}$, where $1 \leq i \leq i' \leq |p|_{row}$, $1 \leq j \leq j' \leq |p|_{col}$, also specified by the pair $[(i, j), (i', j')]$. The portion of p corresponding to positions in subdomain $[(i, j), (i', j')]$ is denoted by $p[(i, j), (i', j')]$. Then a picture x is *subpicture of* p if $x = p[(i, j), (i', j')]$, for some $1 \leq i \leq i' \leq |p|_{row}$, $1 \leq j \leq j' \leq |p|_{col}$. Prefixes of pictures are special subpictures. Given pictures x, p, with $|x|_{row} \leq |p|_{row}$ and $|x|_{col} \leq |p|_{col}$, picture x is a *prefix* of p, denoted $x \trianglelefteq p$, if x is a subpicture of p corresponding to its top-left portion, i.e. if $x = p[(1, 1), (|x|_{row}, |x|_{col})]$.

Dealing with pictures, two concatenation products are classically defined. Let $p, q \in \Sigma^{**}$ be pictures of size (m, n) and (m', n'), respectively. The *column* and the *row concatenation* of p and q are defined by horizontally and vertically juxtaposing p and q. They are partial operations, defined only if $m = m'$ and if $n = n'$, respectively. These operations can be extended to define row- and column- concatenations, and *row-* and *column- stars* on languages. We consider another interesting star operation for picture languages, as introduced by D. Simplot in [24], *the tiling star*. The idea is to compose pictures in some way to cover a rectangular area as, for example, in the following figures.

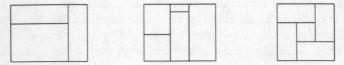

The tiling star of X, denoted by X^{**}, is the set that contains all the empty pictures together with all the non-empty pictures p whose domain can be partitioned in disjoint subdomains $\{d_1, d_2, \ldots, d_k\}$ such that any subpicture p_h of p associated with the subdomain d_h belongs to X, for all $h = 1, \ldots, k$.

Then X^{++} denotes the set X^{**} without the empty pictures. In the sequel, if $p \in X^{++}$, we say that p is *tilable* in X while the partition $t = \{d_1, d_2, \ldots, d_k\}$ of $dom(p)$, together with the corresponding pictures $\{p_1, p_2, \ldots, p_k\}$, is called a *tiling decomposition* of p in X.

3   Two-Dimensional Codes

Let us recall the definitions of codes and strong prefix codes of pictures given in [3,4,7,9], together with some examples. Let Σ be a finite alphabet. $X \subseteq \Sigma^{++}$ is a *code* iff any $p \in \Sigma^{++}$ has at most one tiling decomposition in X.

Example 1. Let $\Sigma = \{a, b\}$ be the alphabet and let $X = \left\{ \boxed{a\ b}, \ \boxed{\begin{smallmatrix}a\\b\end{smallmatrix}}, \ \boxed{\begin{smallmatrix}a\ a\\a\ a\end{smallmatrix}} \right\}$. It is easy to see that X is a code. Any picture $p \in X^{++}$ can be decomposed starting at top-left-corner and checking the subpicture $p[(1,1),(2,2)]$; it can be univocally decomposed in X. Then, proceed similarly for the next contiguous subpictures of size $(2,2)$.

Example 2. Let $X = \left\{ \boxed{a\ b}, \ \boxed{b\ a}, \ \boxed{\begin{smallmatrix}a\\a\end{smallmatrix}} \right\}$. Notice that no picture in X is prefix of another picture in X (see definition in Sect. 2). Nevertheless, X is not a code. Indeed, picture $\boxed{\begin{smallmatrix}a\ b\ a\\a\ b\ a\end{smallmatrix}}$ has the two following different tiling decompositions in X:
$t_1 = \boxed{\begin{smallmatrix}a\ b\ a\\a\ b\ a\end{smallmatrix}}$ and $t_2 = \boxed{\begin{smallmatrix}a\ b\ a\\a\ b\ a\end{smallmatrix}}$.

Taking inspiration from the very remarkable family of prefix codes of strings, let us introduce *strong prefix codes*, defined in [3,9]. The idea is that, given a strong prefix set of pictures $X \subset \Sigma^{++}$, each picture in Σ^{++} can "start" with at most one of the pictures in X.

Definition 3. *Let* $p, q \in \Sigma^{++}$. *Pictures p and q prefix-overlap if for any $(i, j) \in dom(p) \cap dom(q)$, $p(i, j) = q(i, j)$. Moreover pictures p and q strictly prefix-overlap if they prefix-overlap, but neither $p \unlhd q$ nor $q \unlhd p$.*

For example, in the following figure, picture p and q strictly prefix-overlap:

$$\begin{array}{|cc|}\hline a & b \\ a & a \\\hline\end{array} \qquad \begin{array}{|cccc|}\hline a & b & a & a \\\hline\end{array} \qquad \begin{array}{|cc|cc|}\hline a & b & a & a \\ a & a & & \\\hline\end{array}$$

$$p \hspace{5.5em} q \hspace{5.5em} p \text{ and } q \text{ prefix-overlap}$$

Definition 4. *Let $X \subseteq \Sigma^{++}$. X is strong prefix if for any pictures p, q in X with $p \neq q$, p and q do not prefix-overlap.*

Example 5. The following language X is strong prefix; no two pictures in X prefix-overlap.

$$X = \left\{ \boxed{a\ b\ a}\ , \boxed{a\ b\ b}\ , \begin{array}{|c|}\hline b \\ b \\\hline\end{array} , \begin{array}{|cc|}\hline a & a \\ a & a \\\hline\end{array} , \begin{array}{|cc|}\hline a & a \\ a & b \\\hline\end{array} , \begin{array}{|cc|}\hline a & a \\ b & a \\\hline\end{array} , \begin{array}{|cc|}\hline a & a \\ b & b \\\hline\end{array} , \begin{array}{|cc|}\hline b & a \\ a & a \\\hline\end{array} , \begin{array}{|cc|}\hline b & a \\ a & b \\\hline\end{array} , \begin{array}{|cc|}\hline b & b \\ a & a \\\hline\end{array} , \begin{array}{|cc|}\hline b & b \\ a & b \\\hline\end{array} \right\}.$$

Definition 6. *A strong prefix set $X \subseteq \Sigma^{++}$ is maximal strong prefix over Σ if it is not properly contained in any other strong prefix set over Σ; that is, $X \subseteq Y \subseteq \Sigma^{++}$ and Y strong prefix imply $X = Y$.*

The results in [5,9] prove that finite strong prefix codes have a recursive structure and describe an effective procedure to construct all (maximal) finite strong prefix codes of pictures, starting from the "singleton" pictures containing only one alphabet symbol. The construction in some sense extends the literal representation of prefix codes of strings and is based on the notion of *extensions* of a picture. The set of extensions of a picture p to some bigger size (m, n), is the set of all pictures of fixed size (m, n), obtained by adding some columns to the right and some rows to the bottom of p filled with all possible combinations of alphabet symbols.

Let us fix an order between pairs of integers. We write $(m, n) < (m', n')$ if $m \leq m'$, $n \leq n'$ and $m \neq m'$ or $n \neq n'$.

Definition 7. *Let Σ be an alphabet, $p \in \Sigma^{++}$, $m, n \geq 0$ be positive integers with $size(p) < (m, n)$. The set of extensions of p to size (m, n) is $E_{(m,n)}(p) = \{q \in \Sigma^{m,n} \mid q[(1,1), (|p|_{row}, |p|_{col})] = p\}$.*

In [9] the finite maximal strong prefix codes are characterized as follows.

Proposition 8. *$X \subseteq \Sigma^{++}$ is a finite maximal strong prefix code if and only if there exists a finite sequence of picture languages over Σ, $X_1, X_2, \ldots, X_k$, such that $X_1 = \Sigma$, $X = X_k$, and for $i = 1, \ldots, k-1$, $X_{i+1} = (X_i \setminus \{p_i\}) \cup E_{(m_i, n_i)}(p_i)$, for some $p_i \in X_i$, $m_i, n_i \geq 0$.*

4   Infinite Strong Prefix Codes

In this section we consider the strong prefix codes introduced in [3,9] and recalled in the previous Sect. 3. We define a construction for infinite maximal strong prefix codes that provides an interesting inside view of their structure.

We first observe that, as in the one-dimensional case, any strong prefix code of pictures can be embedded into a maximal one. This result allows to concentrate our attention on the infinite strong prefix codes that are maximal.

Proposition 9. *Any strong prefix code $X \subseteq \Sigma^{++}$ is contained in some maximal strong prefix code over Σ.*

The proof is similar to the corresponding one in the one dimensional case (Proposition 1.5 in [14]). It considers, given a strong prefix code X, a chain of strong prefix codes containing X, ordered by set inclusion, and uses the remark that, in view of Zorn's lemma, this chain admits a least upper bound. We omit here all the details.

The following is a simple example of an infinite picture language that is a strong prefix code.

Example 10. Let X be the language of square pictures over $\Sigma = \{a, b\}$ that contains b in all positions apart for the bottom-right corner where symbol a occurs.

$$X = \left\{ \boxed{a} , \begin{array}{|cc|} \hline b & b \\ b & a \\ \hline \end{array} , \begin{array}{|ccc|} \hline b & b & b \\ b & b & b \\ b & b & a \\ \hline \end{array} , \begin{array}{|cccc|} \hline b & b & b & b \\ b & b & b & b \\ b & b & b & b \\ b & b & b & a \\ \hline \end{array} , , \cdots \right\}$$

X is an infinite strong prefix code. Furthermore, X is not maximal strong prefix.

Indeed, consider, for example, the picture $p = \begin{array}{|ccc|} \hline b & b & a \\ b & b & a \\ a & a & a \\ \hline \end{array}$; it is easy to see that $X \cup \{p\}$

is still strong prefix.

Note that X can be viewed as a generalization to 2D of the well known infinite code of strings $S = \{b^n a,\ n \geq 0\}$.

The following example provides a maximal strong prefix code.

Example 11. The language X_∞ contains all square pictures over $\Sigma = \{a, b\}$ such that if p has size (n, n), its prefix of size $(n - 1, n - 1)$ contains only b's while there should be at least one a in the bottom row or in the rightmost column. Then,

$$X_\infty = \left\{ \boxed{a} , \begin{array}{|cc|} \hline b & a \\ a & a \\ \hline \end{array} , \begin{array}{|cc|} \hline b & a \\ a & b \\ \hline \end{array} , \begin{array}{|cc|} \hline b & a \\ b & a \\ \hline \end{array} , \begin{array}{|cc|} \hline b & a \\ b & b \\ \hline \end{array} , \begin{array}{|cc|} \hline b & b \\ a & a \\ \hline \end{array} , \begin{array}{|cc|} \hline b & b \\ a & b \\ \hline \end{array} , \begin{array}{|cc|} \hline b & b \\ b & a \\ \hline \end{array} , \begin{array}{|ccc|} \hline b & b & a \\ b & b & a \\ a & a & a \\ \hline \end{array} , \begin{array}{|ccc|} \hline b & b & b \\ b & b & a \\ a & a & a \\ \hline \end{array} , \begin{array}{|ccc|} \hline b & b & b \\ b & b & b \\ a & a & a \\ \hline \end{array} , \cdots \right\} .$$

The language X_∞ is an infinite maximal strong prefix code. It is immediate to see that it is strong prefix. Indeed, by definition, no picture in X_∞ is prefix of another picture in X_∞ and this implies, since they are all square pictures, that no pair of pictures in X_∞ can prefix-overlap.

To prove the maximality, consider a picture $p \in \Sigma^{++} \setminus X_\infty$. It cannot be $p(1,1) = a$ otherwise $a \in X_\infty$ is a prefix of p. Assume therefore that $p(1,1) = b$. Two cases arise: either $p \in \{b\}^{++}$ or not. In the first case p is a prefix of an infinite number of pictures of X_∞. In the second case, let $b^{k,k}$ be the prefix of p with maximal k, and $k < |p|_{row}, |p|_{col}$. Then there exists a picture $q \in X_\infty$ of size $(k+1, k+1)$, that is a prefix of p. In both cases $X_\infty \cup \{p\}$ is not strong prefix.

Note that the language X_∞ of the previous example contains the language X of the Example 10 (as already noted, X was not maximal strong prefix).

The language X_∞ can be viewed inside a more general family that is obtained by means of *iterated extensions*; the definition takes as starting point the construction of finite maximal strong prefix codes recalled in Proposition 8.

We use the notion of extension of a picture (see Definition 7) to define infinite languages that will result to be maximal strong prefix codes. We give first an informal description. The idea is to construct a language X as infinite union of sets X_k. We start from the initial set $Y_0 = \Sigma$ of all pictures of size $(1,1)$. Then we partition $Y_0 = X_1 \cup A_0$ where X_1 is added to X, while the pictures in A_0 will be extended to get a set of pictures of bigger size. Let Y_1 be the union of all possible extensions of pictures $p \in A_0$ to a size $(m(p), n(p))$ that depends on p. Again we partition $Y_1 = X_2 \cup A_1$ and again we add X_2 to X and take all pictures in A_1 for new extensions to produce the set Y_2. And so on. A further condition ensures that whenever a picture $p \in Y_k$ is not chosen to belong to X_{k+1} (i.e. p stays in A_k to be extended and put in Y_{k+1}), then in some future step, one of its extensions will be surely added to some X_i. Such condition will be crucial in the proof of maximality of Proposition 15. Here below is the formal definition.

Definition 12. *Let Σ be a finite alphabet. A language $X \subseteq \Sigma^{++}$ is generated by* iterated extensions on Σ *if $X = \cup_{k \geq 1} X_k$ where, for any $k \geq 0$,*

(1) $Y_0 = \Sigma$
(2) $A_k \subseteq Y_k$, $X_{k+1} = Y_k \setminus A_k$
(3) $Y_{k+1} = \bigcup_{p \in A_k} E_{(m(p),n(p))}(p)$, for some $(m(p), n(p)) > size(p)$
(4) for any $p \in A_k$, there exist $h > k$ and some extension q of p, with $q \in X_h$.
   The family of all languages generated by iterated extensions on Σ will be denoted by $\mathcal{I}(\pm)$, or simply $\mathcal{I}$, when no ambiguity is possible.

Example 13. The language X_∞ introduced in Example 11 is in $\mathcal{I}$. In fact, $X_\infty = \cup_{k \geq 1} X_k$, where $Y_0 = \Sigma$, $A_0 = \{b\}$, and for any $k \geq 1$, $X_k = Y_{k-1} \setminus A_{k-1}$, with $A_{k-1} = \{p_k\}$ where p_k is the picture of size (k,k) composed of all b's, and $Y_{k-1} = E_{(k,k)}(p_{k-1})$.

Many different and involved languages can be defined by using Definition 12. The matter is to fix the rule to "extract" the set X_{k+1} from Y_k and the criterion to choose the size of the extensions of the pictures in A_k. Consider as an example, the following language.

Example 14. Use iterated extensions on $Y_0 = \{a, b\}$ and take $X_1 = \{a\}$ ($A_0 = \{b\}$) and $Y_1 = E_{(2,2)}(b)$. For any $k \geq 1$, put in X_{k+1} those pictures of Y_k that have the first column equal to the last one. The remaining pictures $p \in Y_k$ (actually pictures of the set A_k) are extended in two different ways. If the last row of p contains an even number of a, add a row to p; if it contains an odd number of a, add a column. This will generate the next set Y_{k+1} containing pictures of many different sizes. Here below, we calculate some of the pictures.

$$X = \left\{ \boxed{a}, \begin{array}{|cc|} \hline b & b \\ a & a \\ \hline \end{array}, \begin{array}{|cc|} \hline b & b \\ b & b \\ \hline \end{array}, \begin{array}{|ccc|} \hline b & a & b \\ a & b & a \\ \hline \end{array}, \begin{array}{|ccc|} \hline b & b & b \\ a & b & a \\ \hline \end{array}, \begin{array}{|ccc|} \hline b & b & b \\ b & a & b \\ \hline \end{array}, \begin{array}{|ccc|} \hline b & a & b \\ a & a & a \\ b & b & b \\ \hline \end{array}, \begin{array}{|ccc|} \hline b & b & b \\ a & a & a \\ b & b & b \\ \hline \end{array}, \dots \right\}.$$

Note that in Definition 12 if, for some $k \geq 0$, $A_k = \emptyset$, then $Y_{k+1}, X_{k+1} = \emptyset$ and the language X is finite. This is the unique case where X can be finite. Otherwise, if for any $k \geq 0$, $A_k \neq \emptyset$, then condition (4) in Definition 12 guarantees that the language is infinite. Moreover, we will see in the next proposition, that condition (4) will be crucial also in proving the maximality of the obtained language.

On the other hand, observe that for some $k \geq 0$, it can hold that $A_k = Y_k$, that is $X_{k+1} = \emptyset$ (without forcing the finiteness of the language).

Next proposition shows that any language generated by iterated extensions is a maximal strong prefix code.

Proposition 15. *Any set $X \in \mathcal{I}(\Sigma)$ is a maximal strong prefix code over Σ.*

Proof. Let $X \in \mathcal{I}(\Sigma)$. First of all, let us show by induction that, for any $h \geq 1$, $\left(\bigcup_{i=1\dots h} X_i \right) \cup A_{h-1}$ is a finite maximal strong prefix code. In the base case, $h = 1$, we have $X_1 \cup A_0 = \{a, b\}$ and this is a maximal strong prefix code. Inductively, suppose that the set $Z = \bigcup_{i=1\dots h-1} X_i \cup A_{h-2}$ is a maximal strong prefix code. Note that the set $\bigcup_{i=1\dots h} X_i \cup A_{h-1}$ can be obtained from Z, by replacing any $p \in A_{h-2} \subseteq Z$ with the set of all its extensions to some bigger size. Hence, it is a finite maximal strong prefix code (see the characterization in Proposition 8).

To show that X is a strong prefix code consider two pictures $p, q \in X$ and suppose $p \in X_h$, $q \in X_k$ and $h \geq k$. Then $p, q \in \bigcup_{i=1\dots h} X_i$ and, since $\bigcup_{i=1\dots h} X_i \cup A_{h-1}$ is a strong prefix code, p and q cannot prefix-overlap.

Now, let us show that X is a maximal strong prefix code. Suppose by contradiction that there exists a picture $p \in \Sigma^{**} \setminus X$ such that $X \cup \{p\}$ is strong prefix. Let $size(p) = (m, n)$ and set $K = max\{k \mid \forall x \in \bigcup_{i=1\dots k} X_i, |x|_{row} \leq m$ and $|x|_{col} \leq n\}$. Consider the set $T = \bigcup_{i=1\dots K} X_i \cup A_{K-1}$. We have $p \notin \bigcup_{i=1\dots k} X_i$ (since $p \notin X$) and $p \notin A_{K-1}$ (since, if $p \in A_{K-1}$, then, by condition *4)* there exists an extension of p in X, against $X \cup \{p\}$ strong prefix). Therefore, $p \notin T$. Let us show that $T \cup \{p\}$ is a strong prefix code, against the maximality of T. Note that p cannot prefix-overlap a picture in $\bigcup_{i=1\dots K} X_i$, since $X \cup \{p\}$ is strong prefix. Furthermore, p cannot strictly prefix-overlap a picture in A_{K-1}. Indeed, any $q \in A_{K-1}$ has a size less than $size(p)$; hence, p and q could strictly prefix-overlap only if q is a prefix of p. This would imply that, there exist some $K' > K$ and some $p' \in X_{K'} \subseteq X$, p' extension of q, such that p' and p prefix-overlap, against $X \cup \{p\}$ strong prefix. We can conclude that $T \cup \{p\}$ is a strong prefix code against the maximality of T. □

We now show the reverse of Proposition 15, i.e. that any maximal infinite strong prefix code can be obtained by iterated extensions.

Proposition 16. *If X is a maximal strong prefix code over Σ then $X \in \mathcal{I}(\Sigma)$.*

Proof (Sketch). Let $Y_0 = \Sigma$, $X_1 = X \cap \Sigma$ and $A_0 = Y_0 \setminus X_1$. The proof is sketched only in the case $\Sigma = \{a, b\}$ and $X_1 = \{b\}$; a similar proof can be used in the other cases. Let us show how to construct the sets X_2, X_3, ..., and so on.

Denote $r_a = \min\{|p|_{row} \mid p \in X \text{ and } a \trianglelefteq p\}$ and $c_a = \min\{|p|_{col} \mid p \in X$ and $a \trianglelefteq p\}$. Clearly $(r_a, c_a) \neq (1,1)$. Set $Y_1 = E_{(r_a, c_a)}(a)$ and $X_2 = X \cap Y_1$; then $A_1 = Y_1 \setminus X_2$. Note that it could be $X_2 = \emptyset$. Observe that the pictures in $X \setminus (X_1 \cup X_2)$ must be the extensions of some pictures in A_1. Indeed, they cannot have a size smaller than the elements in A_1 (for the choice of r_a and c_a); moreover, A_1 contains all pictures in Σ^{r_a, c_a}, except those pictures that are in $X_1 \cup X_2$ (whose extensions cannot be in X, since it is strong prefix). Subsequently, for any $t_1 \in A_1$, at least one extension of t_1 is in X, otherwise the set $X \cup \{t_1\}$ would be strong prefix, against the maximality of X.

For any $q \in A_1$, let $r_q = \min\{|p|_{row} \mid p \in X \text{ and } q \trianglelefteq p\}$ and $c_q = \min\{|p|_{col} \mid p \in X$ and $q \trianglelefteq p\}$. Clearly, $r_q > |q|_{row}$ or $c_q > |q|_{col}$. Set $Y_2 = \bigcup_{q \in A_1} E_{(r_q, c_q)}(q)$, $X_3 = X \cap Y_2$ and $A_2 = Y_2 \setminus X_3$. Again, for any $t_2 \in A_2$, at least one extension of t_2 is in X, otherwise the set $X \cup \{t_2\}$ would be strong prefix. Iterating this scheme, one obtains all the subsequent X_k such that $X = \cup_{k \geq 1} X_k$. □

The results in the two previous propositions can be summarized in the following theorem which gives a characterization of maximal strong prefix codes of pictures. It holds both for finite and infinite codes.

Theorem 17. *Let $X \subseteq \Sigma^{++}$. X is a maximal strong prefix code over Σ if and only if $X \in \mathcal{I}(\Sigma)$.*

5   Measure of Two-Dimensional Languages and Codes

Some important results on codes of strings deal with the notion of measure (cf. [14]). A probability is assigned to each symbol of the alphabet and, for a given string, one multiplies the probability of each letter. Then, the measure of a language is simply the sum of the probability of its strings. A major result states that the measure of a string code is always less than or equal to 1, whereas a thin string code is maximal if and only if its measure is 1. Roughly speaking, a set of strings is not a code if there are "too many too short strings". In this section, we consider the measure of infinite strong prefix codes of pictures as introduced in [5].

Definition 18. *Let Σ be an alphabet and π be a probability distribution on Σ. The* probability *of a picture $p \in \Sigma^{++}$ is defined as $\pi(p) = \prod_{1 \le i \le m, 1 \le j \le n} \pi(p(i,j))$. The* measure *of a language $X \subseteq \Sigma^{++}$ relative to π is $\mu_\pi(X) = \sum_{p \in X} \pi(p)$.*

Particular interest is devoted to the *uniform distribution*, which associates to every symbol a in the alphabet Σ of cardinality k, the probability $\pi_u(a) = \frac{1}{k}$. Then, the *uniform probability* of a picture $p \in \Sigma^{++}$ is $\pi_u(p) = \frac{1}{k^{area(p)}}$. The *uniform measure* of a language $X \subseteq \Sigma^{++}$, is $\mu_u(X) = \sum_{p \in X} \pi_u(p)$.

Example 19. Let $\Sigma = \{a, b\}$ and consider language $X = \left\{ \boxed{b\,b}, \begin{array}{c}\boxed{\begin{array}{c}a\\b\end{array}}\end{array}, \boxed{\begin{array}{cc}a&a\\a&a\end{array}}, \boxed{\begin{array}{cc}a&a\\a&b\end{array}} \right\}$ on Σ. Its uniform measure is $\mu_u(X) = 5/8 < 1$. In general for any probability distribution $\pi(a) = p$, $\pi(b) = 1 - p$, $0 < p < 1$, then $\mu_\pi(X) = p^3 - p + 1 < 1$. Note that X is a code.

A main result in [5] shows that for any finite strong prefix code $X \subseteq \Sigma^{++}$ and measure μ, we have that $\mu(X) \le 1$. Moreover $\mu(X) = 1$ if and only if X is a finite maximal strong prefix code. We show that without the finiteness hypotesis the scenario is different. Coherently with the intuitive relation between code and measure, we prove first the following result.

Theorem 20. *Let $X \subseteq \Sigma^{++}$ be a maximal strong prefix code and μ be a measure. Then $\mu(X) \le 1$.*

Proof. By Theorem 17, and following the notation of Definition 12, X is the union of some languages X_i, for $i \ge 1$. Since the languages X_i's are pairwise disjoint, taking $s_n = \sum_{i=1}^n \mu(X_i)$, we can write $\mu(X) = \lim_{n \to \infty} s_n$. Consider now, for any $n \ge 1$, the sets $Z_n = \bigcup_{i=1...n} X_i \cup A_{n-1}$. For any $n \ge 1$, Z_n is a finite maximal strong prefix code (as shown in the proof of Proposition 15) and therefore $\mu(Z_n) = 1$. Hence, $s_n \le \mu(Z_n) = 1$. Finally, $\mu(X) = \lim s_n \le 1$. $\square$

The measure of infinite maximal strong prefix codes does not behave as the measure of the finite ones. To show this, we propose another example.

Example 21. Consider the language Z_∞ over $\Sigma = \{a, b\}$ that contains the size $(1, 1)$ picture with a and all square pictures p that have symbol b in the top-left position and in all positions of the bottom row and of the rightmost column. Moreover all square prefixes of p should have at least one a in their bottom row or last column,

$$Z_\infty = \left\{ \boxed{a}, \boxed{\begin{array}{c}b\,b\\b\,b\end{array}}, \boxed{\begin{array}{ccc}b&a&b\\a&a&b\\b&b&b\end{array}}, \boxed{\begin{array}{ccc}b&b&b\\a&a&b\\b&b&b\end{array}}, \boxed{\begin{array}{ccc}b&a&b\\b&a&b\\b&b&b\end{array}}, \ldots, \boxed{\begin{array}{cccc}b&a&a&b\\a&a&a&b\\a&a&a&b\\b&b&b&b\end{array}}, \boxed{\begin{array}{cccc}b&a&b&b\\a&a&a&b\\a&a&a&b\\b&b&b&b\end{array}}, \boxed{\begin{array}{cccc}b&b&b&b\\a&a&a&b\\a&a&a&b\\b&b&b&b\end{array}}, \ldots \right\}.$$

The language Z_∞ can be obtained following Definition 12. $Z_\infty = \cup_{i \ge 1} X_i$, where $Y_0 = \Sigma$, $A_0 = \{b\}$, and for any $i \ge 1$, $X_i = Y_{i-1} \setminus A_{i-1}$, with $A_{i-1} = \{p \in \Sigma^{i,i} \mid p$ has at least one a's in the last row or column$\}$, and $Y_{i-1} = \bigcup_{p \in A_{i-2}} E_{(i,i)}(p)$. Then, by Theorem 17, Z_∞ is a maximal strong prefix code.

Proposition 22. *There exist maximal strong prefix codes whose measure is strictly less than 1.*

Proof. Consider the language Z_∞ together with the languages X_i, Y_{i-1}, A_{i-1} resulting by the associated iterated extensions as defined in Example 21. Let us calculate the uniform measure of Z_∞. Since the languages X_i's are pairwise disjoint, $\mu(Z_\infty) = \sum_{i \geq 1} \mu(X_i)$. We have:

- $\mu(X_1) = 1/2$,
- $\mu(X_2) = 1/2^4$, and
- $\mu(X_i) = \frac{(2^3-1)(2^5-1)\cdots(2^{2(i-1)-1}-1)}{2^{1+3+5+\cdots+(2i-1)}}$, for any $i \geq 3$.

Recall that $X_i \subseteq \Sigma^{i,i}$ and $1 + 3 + 5 + \cdots + (2i-1) = i^2$. Then, for any $i \geq 3$,
$\mu(X_i) \leq \frac{2^3 2^5 \ldots 2^{2(i-1)-1}}{2^{1+3+5+\cdots+(2i-1)}} = \frac{1}{2^{1+(2i-1)}} = \frac{1}{2^{2i}} = \frac{1}{4^i}$.
Hence, $\mu(Z_\infty) \leq 1/2 + 1/2^4 + \sum_{i=3}^{\infty} \frac{1}{4^i} = 1/2 + \sum_{i=2}^{\infty}(\frac{1}{4})^i = 1/2 + \sum_{i=0}^{\infty}(\frac{1}{4})^i - 1 - 1/4 = 4/3 - 3/4 = 7/12$. This shows that $\mu(Z_\infty) < 1$. □

The next Proposition characterizes the maximal strong prefix codes which have measure equal to 1, in terms of the measure of the languages involved in its construction by iterated extensions.

Proposition 23. *Let $X \in \mathcal{I}(\Sigma)$ and let A_n, for any $n \geq 0$, be the corresponding languages. The measure of X is equal to 1 if and only if $\lim_{n \to \infty} \mu(A_n) = 0$.*

Proof. Let X_i, Y_{i-1}, A_{i-1}, for any $i \geq 1$, be the languages involved in the iterated extensions for X as in Definition 12.
    Since the languages X_i's are pairwise disjoint, $\mu(X) = \lim_{n \to \infty} s_n$, where $s_n = \sum_{i=1}^{n} \mu(X_i)$. Observe that, for any $i \geq 1$, $\mu(X_i) = \mu(Y_{i-1}) - \mu(A_{i-1})$ and $\mu(Y_i) = \mu(A_{i-1})$, since Y_i contains all the extensions of all the pictures in A_{i-1}. Therefore, $s_n = (\mu(Y_0) - \mu(A_0)) + (\mu(Y_1) - \mu(A_1)) + \cdots + (\mu(Y_n) - \mu(A_n)) = \mu(Y_0) - \mu(A_n) = 1 - \mu(A_n)$. Finally, $\mu(X) = \lim_{n \to \infty} s_n = 1 - \lim_{n \to \infty} \mu(A_n)$. Hence, $\mu(X) = 1$ if and only if $\lim_{n \to \infty} \mu(A_n) = 0$. □

As an application of the previous proposition we prove the following.

Proposition 24. *There exist maximal strong prefix codes whose measure is exactly 1.*

Proof. We consider the language X_∞ as in Example 13 and we show that the uniform measure $\mu(X_\infty) = 1$. Following the construction by iterated extensions, each set A_n contains a single picture p of size $(n+1, n+1)$ then the measure $\mu(A_n) = 1/2^{(n+1)^2}$ and $\lim_{n \to \infty} 1/2^{(n+1)^2} = 0$. By applying Proposition 23 we complete the proof. □

We conclude the paper by observing that the proofs of Propositions 22 and 24 are based on two languages X_∞ and Z_∞ that have somehow complementary structure with respect to the definition by iterated extensions. Starting from $Y_0 = \{a, b\}$, for both languages we take $X_1 = \{a\}$ and $Y_1 = E_{(2,2)}(b)$. At each step i we use the same criterion to partition the respective current sets Y_i (in one side, the only picture with all b's in the bottom row and in the rightmost column and in the other side, all the remaining ones). Nevertheless, for X_∞ such single picture is put in the set A_i to be extended, while for Z_∞ such picture is the only one which is kept in the code. The difference in the cardinality of the two sides of the partition makes the substantial discrepancy in the calculation of the measure.

References

1. Aigrain, P., Beauquier, D.: Polyomino tilings, cellular automata and codicity. Theoret. Comput. Sci. **147**, 165–180 (1995)
2. Anselmo, M., Giammarresi, D., Madonia, M.: Deterministic and unambiguous families within recognizable two-dimensional languages. Fund. Inform. **98**(2–3), 143–166 (2010)
3. Anselmo, M., Giammarresi, D., Madonia, M.: Strong prefix codes of pictures. In: Muntean, T., Poulakis, D., Rolland, R. (eds.) CAI 2013. LNCS, vol. 8080, pp. 47–59. Springer, Heidelberg (2013). doi:10.1007/978-3-642-40663-8_6
4. Anselmo, M., Giammarresi, D., Madonia, M.: Two dimensional prefix codes of pictures. In: Béal, M.-P., Carton, O. (eds.) DLT 2013. LNCS, vol. 7907, pp. 46–57. Springer, Heidelberg (2013). doi:10.1007/978-3-642-38771-5_6
5. Anselmo, M., Giammarresi, D., Madonia, M.: Structure and measure of a decidable class of two-dimensional codes. In: Dediu, A.-H., Formenti, E., Martín-Vide, C., Truthe, B. (eds.) LATA 2015. LNCS, vol. 8977, pp. 315–327. Springer, Cham (2015). doi:10.1007/978-3-319-15579-1_24
6. Anselmo, M., Giammarresi, D., Madonia, M., Restivo, A.: Unambiguous recognizable two-dimensional languages. ITA **40**(2), 227–294 (2006)
7. Anselmo, M., Giammarresi, D., Madonia, M.: Prefix picture codes: a decidable class of two-dimensional codes. Int. J. Found. Comput. Sci. **25**(8), 1017–1032 (2014)
8. Anselmo, M., Giammarresi, D., Madonia, M.: Picture codes and deciphering delay. Information and Computation (2017, in press)
9. Anselmo, M., Giammarresi, D., Madonia, M.: Structure and properties of strong prefix codes of pictures. Math. Struct. Comput. Sci. **27**(2), 123–142 (2017)
10. Anselmo, M., Jonoska, N., Madonia, M.: Framed versus unframed two-dimensional languages. In: Nielsen, M., Kučera, A., Miltersen, P.B., Palamidessi, C., Tůma, P., Valencia, F. (eds.) SOFSEM 2009. LNCS, vol. 5404, pp. 79–92. Springer, Heidelberg (2009). doi:10.1007/978-3-540-95891-8_11
11. Anselmo, M., Madonia, M.: Two-dimensional comma-free and cylindric codes. Theor. Comput. Sci. **658**, 4–17 (2017)
12. Barcucci, E., Bernini, A., Bilotta, S., Pinzani, R.: Cross-bifix-free sets in two dimensions. Theor. Comput. Sci. **664**, 29–38 (2017)
13. Beauquier, D., Nivat, M.: A codicity undecidable problem in the plane. Theoret. Comp. Sci **303**, 417–430 (2003)
14. Berstel, J., Perrin, D., Reutenauer, C.: Codes and Automata. Cambridge University Press, New York (2009)
15. Blum, M., Hewitt, C.: Automata on a 2-dimensional tape. In: SWAT (FOCS), pp. 155–160 (1967)
16. Bozapalidis, S., Grammatikopoulou, A.: Picture codes. ITA **40**(4), 537–550 (2006)

17. Giammarresi, D., Restivo, A.: Two-dimensional languages. In: Rozenberg, G. (ed.) Handbook of Formal Languages, vol. III, pp. 215–268. Springer, Heidelberg (1997)
18. Grammatikopoulou, A.: Prefix picture sets and picture codes. In: Proceedings of the CAI 2005, pp. 255–268. Aristotle University of Thessaloniki (2005)
19. Kari, J., Salo, V.: A survey on picture-walking automata. In: Kuich, W., Rahonis, G. (eds.) Algebraic Foundations in Computer Science. LNCS, vol. 7020, pp. 183–213. Springer, Heidelberg (2011). doi:10.1007/978-3-642-24897-9_9
20. Kolarz, M., Moczurad, W.: Multiset, set and numerically decipherable codes over directed figures. In: Arumugam, S., Smyth, W.F. (eds.) IWOCA 2012. LNCS, vol. 7643, pp. 224–235. Springer, Heidelberg (2012). doi:10.1007/978-3-642-35926-2_25
21. Moczurad, M., Moczurad, W.: Some open problems in decidability of brick (labelled polyomino) codes. In: Chwa, K.-Y., Munro, J.I.J. (eds.) COCOON 2004. LNCS, vol. 3106, pp. 72–81. Springer, Heidelberg (2004). doi:10.1007/978-3-540-27798-9_10
22. Otto, F., Mráz, F.: Deterministic ordered restarting automata for picture languages. Acta Inf. **52**(7–8), 593–623 (2015)
23. Pradella, M., Cherubini, A., Crespi-Reghizzi, S.: A unifying approach to picture grammars. Inf. Comput. **209**(9), 1246–1267 (2011)
24. Simplot, D.: A characterization of recognizable picture languages by tilings by finite sets. Theoret. Comput. Sci. **218**(2), 297–323 (1991)

Restricted Binary Strings
and Generalized Fibonacci Numbers

Antonio Bernini[⊠]

Dipartimento di Matematica E Informatica "U. Dini",
Università degli Studi di Firenze, Viale G.B. Morgagni 65, 50134 Florence, Italy
antonio.bernini@unifi.it

Abstract. We provide some interesting relations involving k-generalized Fibonacci numbers between the set $F_n^{(k)}$ of length n binary strings avoiding k of consecutive 0's and the set of length n strings avoiding $k + 1$ consecutive 0's and 1's with some more restriction on the first and last letter, via a simple bijection. In the special case $k = 2$ a probably new interpretation of Fibonacci numbers is given.

Moreover, we describe in a combinatorial way the relation between the strings of $F_n^{(k)}$ with an odd numbers of 1's and the ones with an even number of 1's.

Keywords: Generalized Fibonacci numbers · Restricted strings · Consecutive patterns avoidance

1 Introduction

In the paper we mainly faces with binary strings avoiding consecutive patterns, providing some enumerative and constructive properties. The notion of pattern was introduced by Knuth [12] about permutations. Then, it was also absorbed within the context of other combinatorial objects as set partitions [11,14], trees [1,6,8,13], so that the notion of pattern has become one of the most studied in the last decades in Combinatorics. Actually, in our paper we are dealing with consecutive patterns [5,10], which probably are the most useful from an applicative point of view being often related to the Theory of Codes [3,4,7].

Very often, during the study of a particular matter, it is possible to have to deal with some related problems which could be deeper investigated. This is what happened during the writing of the paper [2] about non-overlapping matrices. There, it was very important the analysis of some sets of binary strings with some constraints, constituting the row of the studied matrices. In particular, due to enumerative reasons, we needed to find the cardinality of $B_n(0^k, 1^k)$ (the binary strings of length n avoiding k consecutive 0's and k consecutive 1's),

This work is partially supported by the GNCS - INDAM research project 2017 "Codici di stringhe e matrici non sovrapponibili".

A. Dennunzio et al. (Eds.): AUTOMATA 2017, LNCS 10248, pp. 32–43, 2017.
DOI: 10.1007/978-3-319-58631-1_3

$Z_n(0^k, 1^k)$ (the strings in $B_n(0^k, 1^k)$ ending with 0), and $R_n(0^k, 1^k)$ (the strings in $B_n(0^k, 1^k)$ starting with 1 and ending with 0). We observed a strong similarity between the recurrence relation defining the sequence enumerating $B_n(0^k, 1^k)$ and the one defining the k-*generalized Fibonacci numbers*, and we proved [2], in a very easy analytic way and by induction, that the cardinalities of $B_n(0^k, 1^k)$, $Z_n(0^k, 1^k)$, and $R_n(0^k, 1^k)$ are strictly related to the well-known sequence.

One result of the present paper is an alternative explanation of that interesting link. It is well-known [12] that the set of binary strings having length n and avoiding k consecutive 0's, denoted by $F_n^{(k)}$, is enumerated by the k-generalized Fibonacci numbers. Section 2 is devoted to the definition of a bijection between $Z_n(0^k, 1^k)$ and the set $F_{n-1}^{(k-1)}$, so providing a combinatorial and constructive description for the cardinalities of $Z_n(0^k, 1^k)$, $B_n(0^k, 1^k)$, and $R_n(0^k, 1^k)$. Moreover, this section contains a reading of the famous Fibonacci numbers which is new to the best of our knowledge.

The bijection restricted to $R_n(0^k, 1^k)$ maps its strings in the strings of $F_{n-1}^{(k-1)}$ having an odd numbers of 1's, and a purely analytic argument leads to the intriguing fact that the strings in $F_n^{(k)}$ with an odd number of 1's and the ones with an even number of 1's are equinumerous or differ by one string, depending on n and k. Section 3 presents a construction of the set $F_n(k)$ via generating trees, and it reveals the reason why it happens.

2   A Simple Bijection and Some Applications

Let $F_n^{(k)}$ be the set of length n binary strings avoiding k consecutive 0's, with $k \geq 2$. It is known [12] that, $\forall n \geq 0$, denoting by $|F_n^{(k)}|$ the cardinality of $F_n^{(k)}$,

$$|F_n^{(k)}| = \mathrm{f}_{n+k}^{(k)},$$

where $\mathrm{f}_n^{(k)}$ is the sequence of the k-*generalized Fibonacci numbers* defined by

$$\mathrm{f}_n^{(k)} = \begin{cases} 0 & \text{if } 0 \leq n < k-1 \\ 1 & \text{if } n = k-1 \\ \sum_{i=1}^{k} \mathrm{f}_{n-i}^{(k)} & \text{if } n \geq k. \end{cases}$$

Posing $f_n^{(k)} = \mathrm{f}_{n+k}^{(k)}$ we have, $\forall n \geq 0$,

$$f_n^{(k)} = \begin{cases} 2^n & \text{if } 0 \leq n \leq k-1 \\ \sum_{i=1}^{k} f_{n-i}^{(k)} & \text{if } n \geq k \end{cases} \tag{1}$$

and

$$|F_n^{(k)}| = f_n^{(k)}.$$

We denote by $Z_n(0^k, 1^k)$ the set of length n binary strings ending with 0, avoiding k consecutive 0's and k consecutive 1's (these two patterns in the following will be denoted by 0^k and 1^k, respectively). Let $z_n^{(k)} = |Z_n(0^k, 1^k)|$. It can be proved, by induction [2], that

$$z_n^{(k)} = \begin{cases} 1 & \text{if } n = 0 \\ f_{n-1}^{(k-1)} & \text{if } n \geq 1. \end{cases} \tag{2}$$

We provide a simple bijection φ between $Z_n(0^k, 1^k)$ and $F_{n-1}^{(k-1)}$, so giving a combinatorial interpretation of the above formula.

Definition 1. *Let $u \in Z_n(0^k, 1^k)$, $u = u_1 u_2 \ldots u_n$ and let $v = v_1 v_2 \ldots v_{n-1}$ be a string of length $n - 1$. We define the map φ from $Z_n(0^k, 1^k)$ into the set of binary strings of length $n - 1$ such that $\varphi(u) = v$ where*

$$v_i = \begin{cases} 1 \text{ if } u_i \neq u_{i+1} \\ 0 \text{ if } u_i = u_{i+1}, \end{cases}$$

or, equivalently, $v_i = u_i \ XOR \ u_{i+1}$, for $i = 1, 2, \ldots, n - 1$.

We have the following proposition:

Proposition 1. *The map φ is a bijection between $Z_n(0^k, 1^k)$ and $F_{n-1}^{(k-1)}$.*

Proof. First of all, if $u \in Z_n(0^k, 1^k)$, we show that $\varphi(u) = v \in F_{n-1}^{(k-1)}$. Clearly, v has length $n - 1$ by its definition. Moreover, being $u \in Z_n(0^k, 1^k)$, the string u presents at most $k - 1$ equal consecutive symbols (0 and/or 1) so that the string $v = \varphi(u)$ has at most $k - 2$ consecutive symbols equal to 0. In other words, the string v avoids $k - 1$ consecutive 0's and so $v \in F_{n-1}^{(k-1)}$.

It is not difficult to see that the map φ is an injective function: if $u^{(1)}$ and $u^{(2)}$ are two different strings in $Z_n(0^k, 1^k)$, let j be the greatest index such that $u_j^{(1)} \neq u_j^{(2)}$ (surely $j \neq n$ and $u_{j+1}^{(1)} = u_{j+1}^{(2)}$). Then, the two pairs $u_j^{(1)} u_{j+1}^{(1)}$ and $u_j^{(2)} u_{j+1}^{(2)}$ are mapped in two different symbols by φ, so that $\varphi(u^{(1)}) \neq \varphi(u^{(2)})$.

Since $|Z_n(0^k, 1^k)| = |F_{n-1}^{(k-1)}|$ (see (2)) and φ is injective, then φ is also surjective and the thesis follows. ∎

The inverse of φ is easily seen to be defined by $\varphi^{-1}(v) = u$ with

$$u_i = \begin{cases} 0 & \text{if } i = n \\ u_{i+1} & \text{if } v_i = 0 \\ \bar{u}_{i+1} & \text{if } v_i = 1, \end{cases}$$

for $i = n, n-1, \ldots 3, 2, 1$, where $\bar{u}_j = 1$ if $u_j = 0$ and $\bar{u}_j = 0$ if $u_j = 1$. The string u can be recovered from v by starting from the right side: starting from $u_n = 0$, one entry u_i is equal to the successive one if $v_i = 0$, while if $v_i = 1$, then u_i is the complement of u_{i+1}.

The set of length n binary strings avoiding 0^k and 1^k, denoted by $B_n(0^k, 1^k)$ whose cardinality is defined by $b_n^{(k)}$, is strictly related to the set $Z_n(0^k, 1^k)$. More precisely, from [2], we have that

$$z_n^{(k)} = \begin{cases} 1 & \text{if } n = 0 \\ b_n^{(k)}/2 & \text{if } n \geq 1, \end{cases}$$

so that, using (2), it is

$$b_n^{(k)} = \begin{cases} 1 & \text{if } n = 0 \\ 2f_{n-1}^{(k-1)} & \text{if } n \geq 1. \end{cases} \tag{3}$$

Proposition 1 can be used to provide a combinatorial proof of (2) and (3). Moreover, in the special case $k = 3$, it allows the interpretation, probably new, of Fibonacci sequence as the numbers counting the length n binary strings ending with 0 and avoiding 000 and 111 (3 consecutive 0's and 3 consecutive 1's).

Besides the sets $B_n(0^k, 1^k)$ and $Z_n(0^k, 1^k)$, we also consider the set $R_n(0^k, 1^k)$ of the length n binary strings, starting with 1, ending with 0 and avoiding 0^k and 1^k, for $n \geq 1$. Note that if $k = 2$, then $R_n(0^2, 1^2) = \{(10)^{n/2}\}$ if n is even, while if n is odd, then $R_n(0^2, 1^2) = \emptyset$, therefore we consider $k \geq 3$. Posing $R_0(0^k, 1^k) = \{\lambda\}$ and $r_n^{(k)} = |R_n(0^k, 1^k)|$, it can be proved by induction (see [2]) that

$$r_n^{(k)} = \begin{cases} 1 & \text{if } n = 0 \\ \frac{f_{n-1}^{(k-1)} + d_n^{(k)}}{2} & \text{if } n \geq 1 \end{cases} \tag{4}$$

where

$$d_n^{(k)} = \begin{cases} 1 \text{ if } (n \bmod k) = 0 \\ -1 \text{ if } (n \bmod k) = 1 \\ 0 \text{ if } 2 \leq (n \bmod k) \leq k - 1. \end{cases}$$

If $u \in R_n(0^k, 1^k)$, then $\varphi(u)$ is a string of $F_{n-1}^{(k-1)}$ with an odd number of 1's since, clearly, u has an odd number of consecutive pairs of different bits (01 or 10) which are mapped by φ in the bit 1. Inversely, if v is a string of $F_{n-1}^{(k-1)}$ with an odd number of 1's, then it is easily seen that $\varphi^{-1}(v) \in R_n(0^k, 1^k)$. Denoting by $F_{n,odd}^{(k)}$ the subset of $F_n^{(k)}$ of the strings with an odd number of 1's and by $f_{n,o}^{(k)}$ its cardinality, then $R_n(0^k, 1^k)$ is in bijection with $F_{n-1,odd}^{(k-1)}$ (via φ) and, for $n \geq 1$ and $k \geq 3$, due to the above argument,

$$r_n^{(k)} = f_{n-1,o}^{(k-1)}. \tag{5}$$

With the additional notation of $F_{n,even}^{(k)}$ for the subset of $F_n^{(k)}$ of the strings with an even number of 1's and of $f_{n,e}^{(k)}$ for its cardinality, using (4), and (5) and the trivial fact that $f_{n-1}^{(k-1)} = f_{n-1,o}^{(k-1)} + f_{n-1,e}^{(k-1)}$, we deduce that

$$f_{n-1,o}^{(k-1)} = f_{n-1,e}^{(k-1)} + d_n^{(k)} \quad \text{for } n \geq 1 \text{ and } k \geq 3 , \tag{6}$$

which is clearly equivalent to

$$f_{n,o}^{(k)} = f_{n,e}^{(k)} + d_{n+1}^{(k+1)} \quad \text{for } n \geq 0 \text{ and } k \geq 2 , \tag{7}$$

The aim of the next paragraph is to give a combinatorial description of (7).

3 Number of 1's in the Strings of $F_n^{(k)}$

We recall the construction for the strings in $F_n^{(k)}$: if their length is $n \leq k-1$, then all the binary strings having length n belong to $F_n^{(k)}$, while if $n \geq k$, the strings in $F_n^{(k)}$ can be obtained by appending the prefixes $1, 01, 001, \ldots, 0^{k-1}1$ to all the strings with length $n - 1, n - 2, n - 1, \ldots, n - k$, respectively. In other words, denoting by B_n the length n binary strings and using the notation proposed in [15], we have:

$$F_n^{(k)} = \begin{cases} \{\lambda\} & \text{if } n = 0 \\ B_n & \text{if } 1 \leq n < k \\ 1 \cdot F_{n-1}^{(k)} \cup 01 \cdot F_{n-2}^{(k)} \cup \ldots \cup 0^{k-1}1 \cdot F_{n-k}^{(k)} & \text{if } n \geq k . \end{cases} \tag{8}$$

Note that the binary strings in B_n can be obtained in a recursive way by prepending 0 and 1 to the strings in B_{n-1}, starting from the empty string λ:

$$B_n = \begin{cases} \{\lambda\} & \text{if } n = 0 \\ 0 \cdot B_{n-1} \cup 1 \cdot B_{n-1} & \text{if } n \geq 1. \end{cases} \tag{9}$$

The recursive definition (8) of $F_n^{(k)}$ can also be read by means of a set of generating trees. They are a useful tool which is widely employed in Combinatorics (see for example [9,16,17]). Each node of such trees is a binary string of $F_n^{(k)}$ which generates its children at different levels, as we are going to explain in the following.

Definition 2. *Given $k \geq 2$, let w be a length j binary string belonging to B_j, with $0 \leq j < k$. We define a generating tree T_w as follows:*

- *w is the root of T_w;*
- *w has $j + 1$ children which are strings obtained by adding to the left of w the prefixes $0^{k-1-i}1$, from left to right, with $i = j, j - 1, \ldots, 1, 0$, so that the first child is obtained for $i = j$.*

- each node with length $t \geq k$ has k children which are strings obtained by adding to its left the prefixes $0^{k-1-i}1$ from left to right with $i = k-1, k-2, \ldots, 1, 0$, so that the first child is for $i = k-1$.
- the level of each node v, indicated by $\mathcal{L}(v)$, of T_w which is a child of u, $v = 0^{k-1-i}1u$ for some i, is recursively defined as follows:

$$\mathcal{L}(v) = \begin{cases} j & if\ v = w\ (u\ is\ the\ root) \\ \mathcal{L}(u) + k - i & otherwise. \end{cases}$$

Clearly, the level of a node gives the length of the represented string.

Notice that the children of the root appear only starting from level k, while the children of any other node at level $t \geq k$ appear at each level from $t+1$ to $t+k$. Actually, also the children of a node at level $k-1$ appear in the levels from k to $2k-1$. Therefore, we observe that each node v at level $t \geq k-1$ has k children as described in Definition 2, namely $1v$, $01v$, $001v$, $\ldots$, $0^{k-1}1v$. They can be visualized in Fig. 1 where we have labelled the edges leaving v with 1, 01, $\ldots$, 0^{k-1} (corresponding to the prefixes added on the left of v) from left to right, according to Definition 2.

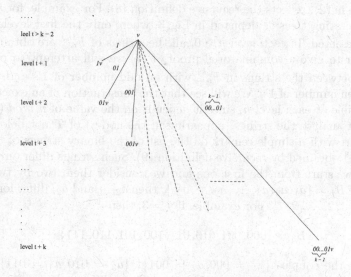

Fig. 1. The children of a node v at level $t \geq k-1$.

In Fig. 2 we present two different generating trees T_w with roots $w = 0$ and $w = 01$, with $k = 4$. In the next figures we omit the labels of the edges.

Considering all the generating trees T_w rooted in each $w \in B_n$ with $0 \leq n < k$, we obtain a set $\mathcal{T}$ of $2^k - 1$ generating trees which are clearly disjoint. It is also straightforward to see that the set $F_n^{(k)}$ coincides with the set of strings appearing in the nodes of the trees of $\mathcal{T}$, since the definition of the children of a

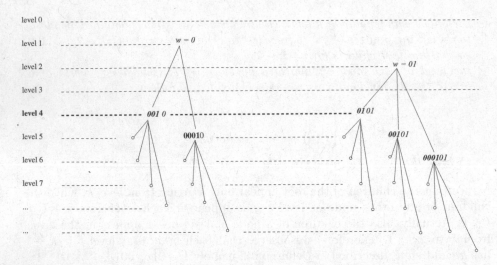

Fig. 2. Two generating trees in the case $k = 4$. The added prefixes are in bold character. Each little circle replaces a string.

node, given in (2), reflects the recursive definition (8). For example, for $k = 3$ we have 7 generating trees as depicted in Fig. 3, where only the first levels of each tree are presented. In each level $n \geq 0$, all the strings of $F_n^{(k)}$ are obtained.

In order to give a combinatorial proof of (7) we will attempt to provide a matching between the strings in $F_n^{(k)}$ with an odd number of 1's and the ones with an even number of 1's. We will see that the construction of an exact pairing is not possible at each level n, since it depends on the value of $n \bmod (k+1)$.

We first analyse the strings (appearing in the nodes) of $\mathcal{T}$ not belonging to $\mathcal{T}_\lambda$. We start with a simple remark on the list of the binary strings of the same length $j \geq 1$ obtained by recursive definition (9). Such strings differ only for the last bit if we start from the first one and we consider them two by two. More precisely, if $B_j = \{a_1, a_2, a_3, \ldots, a_{2^j-1}, a_{2^j}\}$, then a_{2i-1} and a_{2i} differ for the last bit, for $i = 1, 2, \ldots, 2^{j-1}$. For example, if $j = 3$, then

$$B_3 = \{000, 001, 010, 011, 100, 101, 110, 111\}$$

and within the couples $\{a_1 = 000, a_2 = 001\}$, $\{a_3 = 010, a_4 = 011\}$, $\{a_5 = 100, a_6 = 101\}$, $\{a_7 = 110, a_8 = 111\}$ the strings differ only for the last bit. This fact implies evidently that the strings at the same level in the generating trees $T_{a_{2i-1}}$ and $T_{a_{2i}}$ $(i = 1, 2, \ldots, 2^{j-1})$ can be obtained each from each other by switching the last bit, since they are generated by adding the same prefixes to two roots, a_{2i-1} and a_{2i}, differing only in the last bit. In other words, the generating trees $T_{a_{2i-1}}$ and $T_{a_{2i}}$ are isomorphic and two corresponding nodes are obtained by switching the last bit. This is clearly true for each length j of the roots, with $1 \leq j \leq k-1$. Therefore, if at a level n of $\mathcal{T} \setminus \{T_\lambda\}$ there are a certain number of strings in $F_n^{(k)}$ with an odd number of 1's, there are also the

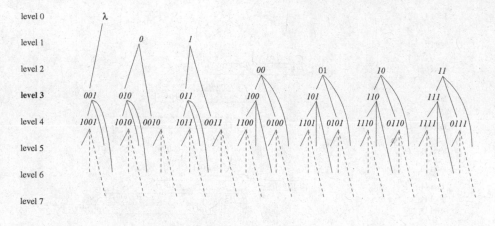

Fig. 3. The seven generating trees in the case $k = 3$. The generated strings are shown only up to level 4. Differently from Fig. 2, we omit the little circles at the end of each edge.

same number of strings with an even number of 1's, at that level. Summarizing, the following proposition holds:

Proposition 2. *The strings in $F_{n,odd}^{(k)}$ and in $F_{n,even}^{(k)}$ not belonging to the generating tree T_λ are equinumerous, for each $n \geq 1$.*

What is left to do in order to combinatorially describe formula (7) is the analysis of the generating tree rooted in the empty string λ. Following Definition 2 the root λ has only one child, namely the string $0^{k-1}1$ at level k, which, on its turn, has k children (see Fig. 4). We denote by $c_i(v) = 0^{i-1}1v$, with $|v| \geq k$ and $i = 1, 2, \ldots, k$, the k children of a node v. Note that $c_1(0^{k-1}1) = 10^{k-1}1$ is the only node at level $k+1$ of the sub-tree $T_{0^{k-1}1}$. Moreover, it can be observed that $c_i(c_1(0^{k-1}1)) = 0^{i-1}110^{k-1}1$ and $c_{i+1}(0^{k-1}1) = 0^i10^{k-1}1$, for $i = 1, 2, \ldots, k-1$, at level $k+1+i$, differ only by the $(k+2)$-th from last digit. Consequently, the sub-trees $T_{c_i(c_1(0^{k-1}1))}$ and $T_{c_{i+1}(0^{k-1}1)}$ are isomorphic and the corresponding strings can be obtained by switching the $(k+2)$-th from last bit (in Fig. 4 isomorphic subtrees have been framed in a rectangular border) . Then, referring to $T_{0^{k-1}1}$ and to Fig. 4, we can summarize that:

1. at level k there is only the string $0^{k-1}1 \in F_{k,odd}^{(k)}$;
2. at level $k+1$ there is only the string $10^{k-1}1 \in F_{k,even}^{(k)}$;
3. from level $k+2$ on, the strings in $F_{n,odd}^{(k)}$ belonging to $T_{c_i(c_1(0^{k-1}1))}$ and $T_{c_{i+1}(0^{k-1}1)}$, for $i = 1, 2, \ldots, k-1$, are as many as the strings in $F_{n,even}^{(k)}$ belonging to the same subtrees, for each $n \geq k+2$.

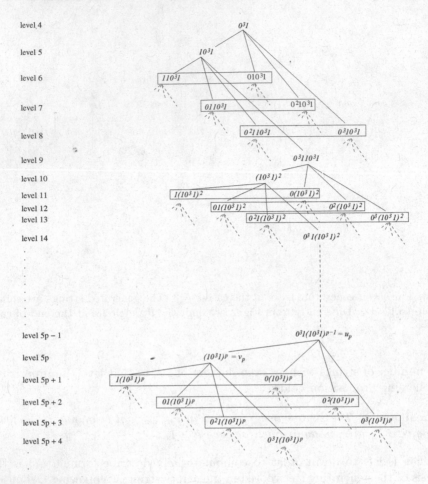

Fig. 4. The subtree $T_{0^{k-1}1}$ in the case $k = 4$. Within each rectangular border there are two isomorphic subtrees.

Note that in point 3 of the above list $i \neq k$, but we have to consider the k-th child of $10^{k-1}1$, i.e. $c_k(10^{k-1}1) = 0^{k-1}110^{k-1}1$, which is the root of a new subtree generating many other strings. We observe that this subtree, $T_{0^{k-1}110^{k-1}1}$, is isomorphic to $T_{0^{k-1}1}1$ and all its nodes can be obtained from the nodes of $T_{0^{k-1}1}$ by appending the suffix $10^{k-1}1$ to each of them. This fact induces a recursive structure on $T_{0^{k-1}1}$. In general, it can be observed that, for any $p \geq 1$, each level $n = (k+1)p - 1$ contains the string $u_p = 0^{k-1}1(10^{k-1}1)^{p-1}$ and each level $n = (k+1)p$ contains the string $v_p = (10^{k-1}1)^p$. Note that u_p and v_p are the only strings which do not belong to any sub-tree rooted in $c_i(v_{p-1})$ or $c_{i+1}(u_{p-1})$, for $i = 1, 2, \ldots, k - 1$ and $p > 1$. Moreover, the strings $c_i(v_p)$ and $c_{i+1}(u_p)$ for $i = 1, 2, \ldots, k - 1$ at level $(k+1)p + i$ differ only for the $((k+1)p + 1)$-th by last bit. Consequently, the sub-trees $T_{c_i(v_p)}$ and $T_{c_{i+1}(u_p)}$ are isomorphic and the corresponding strings can be obtained by switching the $((k+1)p + 1)$-th from last bit. At any level $n \geq (k+1)p+i$, the number of strings with an odd number of 1's deriving from the root $c_i(v_p)$ is equal to the number of strings with an even number of 1's deriving from the root $c_{i+1}(u_p)$, for any $p \geq 1$ and $i = 1, 2, \ldots, k-1$.

Since, evidently, $u_p \in F_{n,odd}^{(k)}$ and $v_p \in F_{n,even}^{(k)}$, we can more generally rephrase the above numbered list in the following proposition:

Proposition 3. *The strings in $F_{n,odd}^{(k)}$ and in $F_{n,even}^{(k)}$ belonging to the generating tree $T_{0^{k-1}1}$ are such that, for any $p \geq 1$,*

- *at each level $n = (k+1)p-1$ the set $F_{n,odd}^{(k)}$ has one string more than $F_{n,even}^{(k)}$, namely $u_p = 0^{k-1}1(10^{k-1}1)^{p-1}$;*
- *at each level $n = (k+1)p$ the set $F_{n,even}^{(k)}$ has one string more than $F_{n,odd}^{(k)}$, namely $v_p = (10^{k-1}1)^p$;*
- *from level $(k+1)p+1$ on, the strings in $F_{n,odd}^{(k)}$ belonging to the subtrees $T_{c_i(v_p)}$ and $T_{c_{i+1}(u_p)}$, for $i = 1, 2, \ldots, k-1$, are as many as the strings in $F_{n,even}^{(k)}$ belonging to the same subtrees.*

Recalling that $|F_{n,odd}^{(k)}| = f_{n,o}^{(k)}$ and $|F_{n,even}^{(k)}| = f_{n,e}^{(k)}$, Propositions 2 and 3 ensure that there is an imbalance between $f_{n,o}^{(k)}$ and $f_{n,e}^{(k)}$ only in the case $n = (k+1)p-1$ and $n = (k+1)p$, for any $p \geq 1$. More precisely, noting that if $n = (k+1)p - 1$ then $n \bmod (k+1) = k$,

$$f_{n,o}^{(k)} = f_{n,e}^{(k)} + 1 \quad \text{if} \quad n \bmod (k+1) = k$$

and

$$f_{n,o}^{(k)} = f_{n,e}^{(k)} - 1 \quad \text{if} \quad n \bmod (k+1) = 0.$$

In all the remaining levels, again from Propositions 2 and 3, it is

$$f_{n,o}^{(k)} = f_{n,e}^{(k)}.$$

Defining

$$\bar{d}_n^{(k+1)} = \begin{cases} 1 & \text{if } (n \bmod (k+1)) = k \\ -1 & \text{if } (n \bmod (k+1)) = 0 \\ 0 & \text{otherwise}, \end{cases}$$

we can write

$$f_{n,o}^{(k)} = f_{n,e}^{(k)} + \bar{d}_n^{(k+1)}.$$

which is equivalent to (7) since, clearly, $\bar{d}_n^{(k+1)} = d_{n+1}^{(k+1)}$.

4   Conclusion

Binary unrestricted strings of length n with an odd number of 1's and the ones with an even number of 1's are clearly equinumerous, since, for instance, a string of the first group gives a string of the second one by switching the last bit. If we consider binary strings of length n avoiding the consecutive pattern 0^k, even if the above easy fact is not true anymore, it is natural to expect that the two subsets have almost the same cardinality. In Sect. 3 of the present paper we provided a rigorous proof of this, showing that they are always the same, except that in particular cases depending on the values of n and k.

The simple bijection φ presented in Sect. 2 let to find a combinatorial and constructive explanation of the cardinalities of the sets $Z_n^{(k)}$ and $R_n^{(k)}$ which formerly were proved only by induction. Moreover, bijection φ led to a probably new interpretation of the famous Fibonacci numbers.

References

1. Bacher, A., Bernini, A., Ferrari, L., Gunby, B., Pinzani, R., West, J.: The Dyck pattern poset. Discrete Math. **321**, 12–23 (2014)
2. Barcucci, E., Bernini, A., Bilotta, S., Pinzani, R.: Non-overlapping matrices. Theoret. Comput. Sci. **658**, 36–45 (2017)
3. Bernini, A., Bilotta, S., Pinzani, R., Sabri, A., Vajnovszki, V.: Prefix partitioned gray codes for particular cross-bifix-free sets. Crypt. Commun. **6**, 359–369 (2014)
4. Bernini, A., Bilotta, S., Pinzani, R., Vajnovszki, V.: A gray code for cross-bifix-free sets. Math. Struct. Comput. Sci. **27**, 184–196 (2017)
5. Bernini, A., Ferrari, L., Steíngrimsson, E.: The Möbius function of the consecutive pattern poset. Electron. J. Comb. **18**(1), #P146 (2011)
6. Bernini, A., Ferrari, L., Pinzani, R., West, J.: Pattern-avoiding Dyck paths. In: Discrete Mathematics and Theoretical Computer Science, vol. AS, pp. 683–694 (2013)
7. Chee, Y.M., Kiah, H.M., Purkayastha, P., Wang, C.: Cross-bifix-free codes within a constant factor of optimality. IEEE Trans. Inf. Theory **59**, 4668–4674 (2013)
8. Dairyko, M., Pudwell, L., Tyner, S., Wynn, C.: Non-contiguous pattern avoidance in binary trees. Electron. J. Comb. **19**(3), #P22 (2012)
9. Elizalde, S.: Generating trees for permutations avoiding generalized patterns. Ann. Comb. **11**, 435–458 (2007)
10. Elizalde, S., Noy, M.: Consecutive patterns in permutations. Adv. Appl. Math. **30**, 110–125 (2003)
11. Klazar, M.: On abab-free and abba-free sets partitions. Eur. J. Comb. **17**, 53–68 (1996)
12. Knuth, D.E.: The Art of Computer Programming: Sorting and Searching, vol. 3. Addison-Wesley, Boston (1966)
13. Rowland, E.: Pattern avoidance in binary trees. J. Comb. Theo. Ser. A **117**, 741–758 (2010)
14. Sagan, B.E.: Pattern avoidance in set partitions. Ars Comb. **94**, 79–96 (2010)

15. Vajnovszki, V.: A loopless generation of bitstrings without p consecutive ones. In: Calude, C.S., Dinneen, M.J., Sburlan, S. (eds.) Combinatorics, Computability and Logic. Discrete Mathematics and Theoretical Computer Science, pp. 227–240. Springer, London (2001)
16. West, J.: Generating trees and the Catalan and Schröder numbers. Discrete Math. **146**, 247–262 (1995)
17. West, J.: Generating trees and forbidden subsequences. Discrete Math. **157**, 363–374 (1996)

Von Neumann Regular Cellular Automata

Alonso Castillo-Ramirez[1($\boxtimes$)] and Maximilien Gadouleau[2]

[1] Departamento de Matemáticas, Centro Universitario de Ciencias Exactas
e Ingenierías, Universidad de Guadalajara, Guadalajara, Mexico
alonso.castillor@academicos.udg.mx
[2] School of Engineering and Computing Sciences,
Durham University, South Road, Durham DH1 3LE, UK

Abstract. For any group G and any set A, a cellular automaton (CA) is a transformation of the configuration space A^G defined via a finite memory set and a local function. Let $\mathrm{CA}(G; A)$ be the monoid of all CA over A^G. In this paper, we investigate a generalisation of the inverse of a CA from the semigroup-theoretic perspective. An element $\tau \in \mathrm{CA}(G; A)$ is *von Neumann regular* (or simply *regular*) if there exists $\sigma \in \mathrm{CA}(G; A)$ such that $\tau \circ \sigma \circ \tau = \tau$ and $\sigma \circ \tau \circ \sigma = \sigma$, where $\circ$ is the composition of functions. Such an element σ is called a *generalised inverse* of τ. The monoid $\mathrm{CA}(G; A)$ itself is regular if all its elements are regular. We establish that $\mathrm{CA}(G; A)$ is regular if and only if $|G| = 1$ or $|A| = 1$, and we characterise all regular elements in $\mathrm{CA}(G; A)$ when G and A are both finite. Furthermore, we study regular linear CA when $A = V$ is a vector space over a field $\mathbb{F}$; in particular, we show that every regular linear CA is invertible when G is torsion-free (e.g. when $G = \mathbb{Z}^d, d \geq 1$), and that every linear CA is regular when V is finite-dimensional and G is locally finite with $\mathrm{char}(\mathbb{F}) \nmid o(g)$ for all $g \in G$.

Keywords: Cellular automata · Linear cellular automata · Monoids · von Neumann regular elements · Generalised inverses

1 Introduction

Cellular automata (CA), introduced by John von Neumann and Stanislaw Ulam in the 1940s, are models of computation with important applications to computer science, physics, and theoretical biology. We follow the modern general setting for CA presented in [5]. For any group G and any set A, a CA over G and A is a transformation of the configuration space A^G defined via a finite memory set and a local function. Most of the classical literature on CA focus on the case when $G = \mathbb{Z}^d$, for $d \geq 1$, and A is a finite set (see [11]), but important results have been obtained for larger classes of groups (e.g., see [5] and references therein).

Recall that a *semigroup* is a set equipped with an associative binary operation, and that a *monoid* is a semigroup with an identity element. Let $\mathrm{CA}(G; A)$ be the set of all CA over G and A. It turns out that, equipped with the composition of functions, $\mathrm{CA}(G; A)$ is a monoid. In this paper we apply functions on

A. Dennunzio et al. (Eds.): AUTOMATA 2017, LNCS 10248, pp. 44–55, 2017.
DOI: 10.1007/978-3-319-58631-1_4

the right; hence, for $\tau, \sigma \in \mathrm{CA}(G; A)$, the composition $\tau \circ \sigma$, denoted simply by $\tau\sigma$, means applying first τ and then σ.

In general, $\tau \in \mathrm{CA}(G; A)$ is *invertible*, or *reversible*, or *a unit*, if there exists $\sigma \in \mathrm{CA}(G; A)$ such that $\tau\sigma = \sigma\tau = \mathrm{id}$. In such case, σ is called *the inverse* of τ and denoted by $\sigma = \tau^{-1}$. When A is finite, it may be shown that $\tau \in \mathrm{CA}(G; A)$ is invertible if and only if it is a bijective function (see [5, Theorem 1.10.2]).

We shall consider the notion of *regularity* which, coincidentally, was introduced by John von Neumann in the context of rings, and has been widely studied in semigroup theory (recall that the multiplicative structure of a ring is precisely a semigroup). Intuitively, cellular automaton $\tau \in \mathrm{CA}(G; A)$ is *von Neumann regular* if there exists $\sigma \in \mathrm{CA}(G; A)$ mapping any configuration in the image of τ to one of its preimages under τ. Clearly, this generalises the notion of reversibility.

Henceforth, we use the term 'regular' to mean 'von Neumann regular'. Let S be any semigroup. For $a, b \in S$, we say that b is *a weak generalised inverse* of a if

$$aba = a.$$

We say that b is *a generalised inverse* (often just called *an inverse*) of a if

$$aba = a \text{ and } bab = b.$$

An element $a \in S$ may have none, one, or more (weak) generalised inverses. It is clear that any generalised inverse of a is also a weak generalised inverse; not so obvious is that, given the set $W(a)$ of weak generalised inverses of a we may obtain the set $V(a)$ of generalised inverses of a as follows (see [6, Exercise 1.9.7]):

$$V(a) = \{bab' : b, b' \in W(a)\}.$$

An element $a \in S$ is *regular* if it has at least one generalised inverse (which is equivalent of having at least one weak generalised inverse). A semigroup S itself is called *regular* if all its elements are regular. Many of the well-known types of semigroups are regular, such as idempotent semigroups (or *bands*), full transformation semigroups, and Rees matrix semigroups. Among various advantages, regular semigroups have a particularly manageable structure which may be studied using the so-called Green's relations. For further basic results on regular semigroups see [6, Sect. 1.9].

Another generalisation of reversible CA has appeared in the literature before [13,14] using the concept of *Drazin inverse* [8]. However, as Drazin invertible elements are a special kind of regular elements, our approach turns out to be more general and natural.

In the following sections we study the regular elements in monoids of CA. First, in Sect. 2 we present some basic results and examples, and we establish that, except for the trivial cases $|G| = 1$ and $|A| = 1$, the monoid $\mathrm{CA}(G; A)$ is not regular. In Sect. 3, we study the regular elements of $\mathrm{CA}(G; A)$ when G and A are both finite; in particular, we characterise them and describe a regular submonoid. In Sect. 4, we study the regular elements of the monoid $\mathrm{LCA}(G; V)$ of linear CA, when V is a vector space over a field $\mathbb{F}$. Specifically, using results on group rings, we show that, when G is torsion-free (e.g., $G = \mathbb{Z}^d$), $\tau \in \mathrm{LCA}(G;V)$ is regular if and only if it is invertible, and that, for finite-dimensional V, $\mathrm{LCA}(G;V)$ itself is regular if and only if G is locally finite and $\mathrm{char}(\mathbb{F}) \nmid |\langle g \rangle|$, for all $g \in G$. Finally, for the particular case when $G \cong \mathbb{Z}_n$ is a cyclic group, $V := \mathbb{F}$ is a finite field, and $\mathrm{char}(\mathbb{F}) \mid n$, we count the total number of regular elements in $\mathrm{LCA}(\mathbb{Z}_n; \mathbb{F})$.

2   Regular Cellular Automata

For any set X, let $\mathrm{Tran}(X)$, $\mathrm{Sym}(X)$, and $\mathrm{Sing}(X)$, be the sets of all functions, all bijective functions, and all non-bijective (or singular) functions of the form $\tau : X \to X$, respectively. Equipped with the composition of functions, $\mathrm{Tran}(X)$ is known as the *full transformation monoid* on X, $\mathrm{Sym}(X)$ is the *symmetric group* on X, and $\mathrm{Sing}(X)$ is the *singular transformation semigroup* on X. When X is a finite set of size α, we simply write Tran_α, Sym_α, and Sing_α, in each case.

We shall review the broad definition of CA that appears in [5, Sect. 1.4]. Let G be a group and A a set. Denote by A^G the *configuration space*, i.e. the set of all functions of the form $x : G \to A$. For each $g \in G$, denote by $R_g : G \to G$ the right multiplication function, i.e. $(h)R_g := hg$ for any $h \in G$. We emphasise that we apply functions on the right, while [5] applies functions on the left.

Definition 1. *Let G be a group and A a set. A* cellular automaton *over G and A is a transformation $\tau : A^G \to A^G$ satisfying the following: there is a finite subset $S \subseteq G$, called a* memory set *of τ, and a local function $\mu : A^S \to A$ such that*

$$(g)(x)\tau = ((R_g \circ x)|_S)\mu, \ \forall x \in A^G, g \in G,$$

where $(R_g \circ x)|_S$ is the restriction to S of $(R_g \circ x) : G \to A$.

The group G acts on the configuration space A^G as follows: for each $g \in G$ and $x \in A^G$, the configuration $x \cdot g \in A^G$ is defined by

$$(h)x \cdot g := (hg^{-1})x, \quad \forall h \in G.$$

A transformation $\tau : A^G \to A^G$ is *G-equivariant* if, for all $x \in A^G$, $g \in G$,

$$(x \cdot g)\tau = ((x)\tau) \cdot g.$$

Any cellular automaton is G-equivariant, but the converse is not true in general. A generalisation of Curtis-Hedlund Theorem (see [5, Theorem 1.8.1]) establishes that, when A is finite, $\tau : A^G \to A^G$ is a CA if and only if τ is G-equivariant and continuous in the prodiscrete topology of A^G; in particular, when G and A are both finite, G-equivariance completely characterises CA over G and A.

A configuration $x \in A^G$ is called *constant* if $(g)x = k$, for a fixed $k \in A$, for all $g \in G$. In such case, we denote x by $\mathbf{k} \in A^G$.

Remark 1. It follows by G-equivariance that any $\tau \in \mathrm{CA}(G;A)$ maps constant configurations to constant configurations.

Recall from Sect. 1 that $\tau \in \mathrm{CA}(G; A)$ is *invertible* if there exists $\sigma \in \mathrm{CA}(G; A)$ such that $\tau\sigma = \sigma\tau = \mathrm{id}$, and that $\tau \in \mathrm{CA}(G; A)$ is *regular* if there exists $\sigma \in \mathrm{CA}(G; A)$ such that $\tau\sigma\tau = \tau$. We now present some examples of CA that are regular but not invertible.

Example 1. Let G be any nontrivial group and A any set with at least two elements. Let $\sigma \in \mathrm{CA}(G; A)$ be a CA with memory set $\{s\} \subseteq G$ and local function $\mu : A \to A$ that is non-bijective. Clearly, σ is not invertible. As $\mathrm{Sing}(A)$ is a regular semigroup (see [10, Theorem II]), there exists $\mu' : A \to A$ such that $\mu\mu'\mu = \mu$. If σ' is the CA with memory set $\{s^{-1}\}$ and local function μ', then $\sigma\sigma'\sigma = \sigma$. Hence σ is regular.

Example 2. Suppose that $A = \{0, 1, \ldots, q - 1\}$, with $q \geq 2$. Consider $\tau_1, \tau_2 \in \mathrm{CA}(\mathbb{Z}; A)$ with memory set $S := \{-1, 0, 1\}$ and local functions

$$(x)\mu_1 = \min\{(-1)x, (0)x, (1)x\} \text{ and } (x)\mu_2 = \max\{(-1)x, (0)x, (1)x\},$$

respectively, for all $x \in A^S$. Clearly, τ_1 and τ_2 are not invertible, but we show that they are generalised inverses of each other, i.e. $\tau_1\tau_2\tau_1 = \tau_1$ and $\tau_2\tau_1\tau_2 = \tau_2$, so they are both regular. We prove only the first of the previous identities, as the second one is symmetrical. Let $x \in A^{\mathbb{Z}}$, $y := (x)\tau_1$, $z := (y)\tau_2$, and $a := (z)\tau_1$. We want to show that $y = a$. For all $i \in \mathbb{Z}$ and $\epsilon \in \{-1, 0, 1\}$, we have

$$(i + \epsilon)y = \min\{(i + \epsilon - 1)x, (i + \epsilon)x, (i + \epsilon + 1)x\} \leq (i)x.$$

Hence,

$$(i)z = \max\{(i - 1)y, (i)y, (i + 1)y\} \leq (i)x.$$

Similarly $(i - 1)z \leq (i - 1)x$ and $(i + 1)z \leq (i + 1)x$, so

$$(i)a = \min\{(i - 1)z, (i)z, (i + 1)z\} \leq (i)y = \min\{(i - 1)x, (i)x, (i + 1)x\}.$$

Conversely, we have $(i - 1)z, (i)z, (i + 1)z \geq (i)y$, so $(i)a \geq (i)y$. In particular, when $q = 2$, τ_1 and τ_2 are the elementary CA known as Rules 128 and 254, respectively.

The following lemma gives an equivalent definition of regular CA. Note that this result still holds if we replace $\mathrm{CA}(G; A)$ with any monoid of transformations.

Lemma 1. *Let G be a group and A a set. Then, $\tau \in \mathrm{CA}(G; A)$ is regular if and only if there exists $\sigma \in \mathrm{CA}(G; A)$ such that for every $y \in (A^G)\tau$ there is $\hat{y} \in A^G$ with $(\hat{y})\tau = y$ and $(y)\sigma = \hat{y}$.*

Proof. If $\tau \in \mathrm{CA}(G; A)$ is regular, there exists $\sigma \in \mathrm{CA}(G; A)$ such that $\tau\sigma\tau = \tau$. Let $x \in A^G$ be such that $(x)\tau = y$ (which exists because $y \in (A^G)\tau$) and define $\hat{y} := (y)\sigma$. Now,

$$(\hat{y})\tau = (y)\sigma\tau = (x)\tau\sigma\tau = (x)\tau = y.$$

Conversely, assume there exists $\sigma \in \mathrm{CA}(G; A)$ satisfying the statement of the lemma. Then, for any $x \in A^G$ with $y := (x)\tau$ we have

$$(x)\tau\sigma\tau = (y)\sigma\tau = (\hat{y})\tau = y = (x)\tau.$$

Therefore, τ is regular. $\square$

Corollary 1. *Let G be a nontrivial group and A a set with at least two elements. Let $\tau \in \mathrm{CA}(G; A)$, and suppose there is a constant configuration $\mathbf{k} \in (A^G)\tau$ such that there is no constant configuration of A^G mapped to $\mathbf{k}$ under τ. Then τ is not regular.*

Proof. The result follows by Remark 1 and Lemma 1.   □

In the following examples we see how Corollary 1 may be used to show that some well-known CA are not regular.

Example 3. Let $\phi \in \mathrm{CA}(\mathbb{Z}; \{0, 1\})$ be the Rule 110 elementary CA, and consider the constant configuration $\mathbf{1}$. Define $x := \ldots 10101010 \cdots \in \{0, 1\}^{\mathbb{Z}}$, and note that $(x)\phi = \mathbf{1}$. Since $(\mathbf{1})\phi = \mathbf{0}$ and $(\mathbf{0})\phi = \mathbf{0}$, Corollary 1 implies that ϕ is not regular.

Example 4. Let $\tau \in \mathrm{CA}(\mathbb{Z}^2; \{0, 1\})$ be Conway's Game of Life, and consider the constant configuration $\mathbf{1}$ (all cells alive). By [5, Exercise 1.7.], $\mathbf{1}$ is in the image of τ; since $(\mathbf{1})\tau = \mathbf{0}$ (all cells die from overpopulation) and $(\mathbf{0})\tau = \mathbf{0}$, Corollary 1 implies that τ is not regular.

The following theorem applies to CA over arbitrary groups and sets, and it shows that, except for the trivial cases, $\mathrm{CA}(G; A)$ always contains non-regular elements.

Theorem 1. *Let G be a group and A a set. The semigroup $\mathrm{CA}(G; A)$ is regular if and only if $|G| = 1$ or $|A| = 1$.*

Proof. If $|G| = 1$ or $|A| = 1$, then $\mathrm{CA}(G; A) = \mathrm{Tran}(A)$ or $\mathrm{CA}(G; A)$ is the trivial semigroup with one element, respectively. In both cases, $\mathrm{CA}(G; A)$ is regular (see [6, Exercise 1.9.1]).

Assume that $|G| \geq 2$ and $|A| \geq 2$. Suppose that $\{0, 1\} \subseteq A$. Let $S := \{e, g, g^{-1}\} \subseteq G$, where e is the identity of G and $e \neq g \in G$ (we do not require $g \neq g^{-1}$). For $i = 1, 2$, let $\tau_i \in \mathrm{CA}(G; A)$ be the cellular automaton defined by the local function $\mu_i : A^S \to A$, where, for any $x \in A^S$,

$$(x)\mu_1 := \begin{cases} (e)x & \text{if } (e)x = (g)x = (g^{-1})x, \\ 0 & \text{otherwise;} \end{cases}$$

$$(x)\mu_2 := \begin{cases} 1 & \text{if } (e)x = (g)x = (g^{-1})x = 0, \\ (e)x & \text{otherwise.} \end{cases}$$

We shall show that $\tau := \tau_2 \tau_1 \in \mathrm{CA}(G; A)$ is not regular.

Consider the constant configurations $\mathbf{0}, \mathbf{1} \in A^G$. Let $z \in A^G$ be defined by

$$(h)z := \begin{cases} m \mod (2) & \text{if } h = g^m, m \in \mathbb{N} \text{ minimal}, \\ 0 & \text{otherwise.} \end{cases}$$

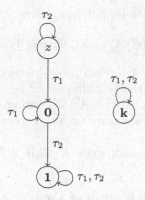

Fig. 1. Images of τ_1 and τ_2.

Figure 1 illustrates the images z, $\mathbf{0}$, $\mathbf{1}$, and $\mathbf{k} \neq \mathbf{0}, \mathbf{1}$ (in case it exists) under τ_1 and τ_2. Clearly,

$$(\mathbf{0})\tau = (\mathbf{0})\tau_2\tau_1 = (\mathbf{1})\tau_1 = \mathbf{1}.$$

In fact,

$$(\mathbf{k})\tau = \begin{cases} \mathbf{1} & \text{if } \mathbf{k} = \mathbf{0}, \\ \mathbf{k} & \text{otherwise.} \end{cases}$$

Furthermore,

$$(z)\tau = (z)\tau_2\tau_1 = (z)\tau_1 = \mathbf{0}.$$

Hence, $\mathbf{0}$ is a constant configuration in the image of τ but with no preimage among the constant configurations. By Corollary 1, τ is not regular.   □

Now that we know that $CA(G; A)$ always contains both regular and non-regular elements (when $|G| \geq 2$ and $|A| \geq 2$), an interesting problem is to find a criterion that describes all regular CA. In the following sections, we solve this problem by adding some extra assumptions, such as finiteness and linearity.

3   Regular Finite Cellular Automata

In this section we characterise the regular elements in the monoid $CA(G; A)$ when G and A are both finite (Theorem 3). In order to achieve this, we summarise some of the notation and results obtained in [2–4].

Definition 2. *The following definitions apply for an arbitrary group G and an arbitrary set A:*

1. *For any $x \in A^G$, the G-orbit of x in A^G is $xG := \{x \cdot g : g \in G\}$.*
2. *For any $x \in A^G$, the stabiliser of x in G is $G_x := \{g \in G : x \cdot g = x\}$.*
3. *A subshift of A^G is a subset $X \subseteq A^G$ that is G-invariant, i.e. for all $x \in X$, $g \in G$, we have $x \cdot g \in X$, and closed in the prodiscrete topology of A^G.*

4. The group of invertible cellular automata *over G and A is*

$$\text{ICA}(G; A) := \{\tau \in \text{CA}(G; A) : \exists \phi \in \text{CA}(G; A) \text{ such that } \tau\phi = \phi\tau = \text{id}\}.$$

In the case when G and A are both finite, every subset of A^G is closed in the prodiscrete topology, so the subshifts of A^G are simply unions of G-orbits. Moreover, as every map $\tau : A^G \to A^G$ is continuous in this case, $\text{CA}(G; A)$ consists of all the G-equivariant maps of A^G. Theorem 2 is easily deduced from Lemmas 3, 9 and 10 in [4].

If M is a group, or a monoid, write $K \leq M$ if K is a subgroup, or a submonoid, of M, respectively.

Theorem 2. *Let G be a finite group of size $n \geq 2$ and A a finite set of size $q \geq 2$. Let $x, y \in A^G$.*

(i) *Let $\tau \in \text{CA}(G; A)$. If $(x)\tau \in (xG)$, then $\tau|_{xG} \in \text{Sym}(xG)$.*
(ii) *There exists $\tau \in \text{ICA}(G; A)$ such that $(x)\tau = y$ if and only if $G_x = G_y$.*
(iii) *There exists $\tau \in \text{CA}(G; A)$ such that $(x)\tau = y$ if and only if $G_x \leq G_y$.*

Theorem 3. *Let G be a finite group and A a finite set of size $q \geq 2$. Let $\tau \in \text{CA}(G; A)$. Then, τ is regular if and only if for every $y \in (A^G)\tau$ there is $x \in A^G$ such that $(x)\tau = y$ and $G_x = G_y$.*

Proof. First, suppose that τ is regular. By Lemma 1, there exists $\phi \in \text{CA}(G; A)$ such that for every $y \in (A^G)\tau$ there is $\hat{y} \in A^G$ with $(\hat{y})\tau = y$ and $(y)\phi = \hat{y}$. Take $x := \hat{y}$. By Theorem 2, $G_x \leq G_y$ and $G_y \leq G_x$. Therefore, $G_x = G_y$.

Conversely, suppose that for every $y \in (A^G)\tau$ there is $x \in A^G$ such that $(x)\tau = y$ and $G_x = G_y$. Choose pairwise distinct G-orbits $y_1 G, \ldots, y_\ell G$ such that

$$(A^G)\tau = \bigcup_{i=1}^{\ell} y_i G.$$

For each i, fix $y_i' \in A^G$ such that $(y_i')\tau = y_i$ and $G_{y_i} = G_{y_i'}$. We define $\phi : A^G \to A^G$ as follows: for any $z \in A^G$,

$$(z)\phi := \begin{cases} z & \text{if } z \notin (A^G)\tau, \\ y_i' \cdot g & \text{if } z = y_i \cdot g \in y_i G. \end{cases}$$

The map ϕ is well-defined because

$$y_i \cdot g = y_i \cdot h \iff hg^{-1} \in G_{y_i} = G_{y_i'} \iff y_i' \cdot g = y_i' \cdot h.$$

Clearly, ϕ is G-equivariant, so $\phi \in \text{CA}(G; A)$. Now, for any $x \in A^G$ with $(x)\tau = y_i \cdot g$,

$$(x)\tau\phi\tau = (y_i \cdot g)\phi\tau = (y_i' \cdot g)\tau = (y_i')\tau \cdot g = y_i \cdot g = (x)\tau.$$

This proves that $\tau\phi\tau = \tau$, so τ is regular. $\qquad\square$

Our goal now is to find a regular submonoid of $\mathrm{CA}(G; A)$ and describe its structure (see Theorem 4). In order to achieve this, we need some further terminology and basic results.

Say that two subgroups H_1 and H_2 of G are *conjugate* in G if there exists $g \in G$ such that $g^{-1}H_1g = H_2$. This defines an equivalence relation on the subgroups of G. Denote by $[H]$ the conjugacy class of $H \leq G$. Define the *box* in A^G corresponding to $[H]$, where $H \leq G$, by

$$B_{[H]}(G; A) := \{x \in A^G : [G_x] = [H]\}.$$

As any subgroup of G is the stabiliser of some configuration in A^G, the set $\{B_{[H]}(G; A) : H \leq G\}$ is a partition of A^G. Note that $B_{[H]}(G; A)$ is a subshift of A^G (because $G_{(x \cdot g)} = g^{-1}G_x g$) and, by the Orbit-Stabiliser Theorem, all the G-orbits contained in $B_{[H]}(G; A)$ have equal sizes. When G and A are clear from the context, we write simply $B_{[H]}$ instead of $B_{[H]}(G; A)$.

Example 5. For any finite group G and finite set A of size q, we have

$$B_{[G]} = \{\mathbf{k} \in A^G : \mathbf{k} \text{ is constant}\}.$$

For any subshift $C \subseteq A^G$, define

$$\mathrm{CA}(C) := \{\tau \in \mathrm{Tran}(C) : \tau \text{ is } G\text{-equivariant}\}.$$

In particular, $\mathrm{CA}(A^G) = \mathrm{CA}(G; A)$. Clearly,

$$\mathrm{CA}(C) = \{\tau|_C : \tau \in \mathrm{CA}(G; A), \tau(C) \subseteq C\}.$$

A submonoid $R \leq M$ is called *maximal regular* if there is no regular monoid K such that $R < K < M$.

Theorem 4. *Let G be a finite group and A a finite set of size $q \geq 2$. Let*

$$R := \left\{\sigma \in \mathrm{CA}(G; A) : G_x = G_{(x)\sigma} \text{ for all } x \in A^G\right\}.$$

(i) $\mathrm{ICA}(G; A) \leq R$.
(ii) R *is a regular monoid.*
(iii) $R \cong \prod_{H \leq G} \mathrm{CA}(B_{[H]})$.
(iv) R *is not a maximal regular submonoid of* $\mathrm{CA}(G; A)$.

Proof. Part **(i)** and **(iii)** are trivial while part **(ii)** follows by Theorem 3.

For part **(iv)**, let $x, y \in A^G$ be such that $G_x < G_y$, so x and y are in different boxes. Define $\tau \in \mathrm{CA}(G; A)$ such that $(x)\tau = y$, $(B_{[G_y]})\tau = yG$, and τ fixes any other configuration in $A^G \backslash (B_{[G_y]} \cup \{xG\})$. It is clear by Theorem 3 that τ is regular. We will show that $K := \langle R, \tau \rangle$ is a regular submonoid of $\mathrm{CA}(G; A)$. Let $\sigma \in K$ and $z \in (A^G)\sigma$. If $\sigma \in R$, then it is obviously regular, so assume that $\sigma = \rho_1 \tau \rho_2$ with $\rho_1 \in K$ and $\rho_2 \in R$. If $z \in A^G \backslash (B_{[G_y]})$, it is clear that z has a preimage in its own box; otherwise $(B_{[G_y]})\sigma = (yG)\rho_2 = zG$ and z has a preimage in $B_{[G_y]}$. Hence σ is regular and so is K. $\square$

4   Regular Linear Cellular Automata

Let V a vector space over a field $\mathbb{F}$. For any group G, the configuration space V^G is also a vector space over $\mathbb{F}$ equipped with the pointwise addition and scalar multiplication. Denote by $\operatorname{End}_{\mathbb{F}}(V^G)$ the set of all $\mathbb{F}$-linear transformations of the form $\tau : V^G \to V^G$. Define

$$\operatorname{LCA}(G; V) := \operatorname{CA}(G; V) \cap \operatorname{End}_{\mathbb{F}}(V^G).$$

Note that $\operatorname{LCA}(G; V)$ is not only a monoid, but also an $\mathbb{F}$-algebra (i.e. a vector space over $\mathbb{F}$ equipped with a bilinear binary product), because, again, we may equip $\operatorname{LCA}(G; V)$ with the pointwise addition and scalar multiplication. In particular, $\operatorname{LCA}(G; V)$ is also a ring.

As in the case of semigroups, von Neumann regular rings have been widely studied and many important results have been obtained. In this chapter, we study the regular elements of $\operatorname{LCA}(G; V)$ under some natural assumptions on the group G.

First, we introduce some preliminary results and notation. The *group ring* $R[G]$ is the set of all functions $f : G \to R$ with finite support (i.e. the set $\{g \in G : (g)f \neq 0\}$ is finite). Equivalently, the group ring $R[G]$ may be defined as the set of all formal finite sums $\sum_{g \in G} a_g g$ with $a_g \in R$. The multiplication in $R[G]$ is defined naturally using the multiplications of G and R:

$$\sum_{g \in G} a_g g \sum_{h \in G} a_h h = \sum_{g,h \in G} a_g a_h gh.$$

If we let $R := \operatorname{End}_{\mathbb{F}}(V)$, it turns out that $\operatorname{End}_{\mathbb{F}}(V)[G]$ is isomorphic to $\operatorname{LCA}(G; V)$ as $\mathbb{F}$-algebras (see [5, Theorem 8.5.2]).

Define the *order* of $g \in G$ by $o(g) := |\langle g \rangle|$ (i.e. the size of the subgroup generated by g). The group G is *torsion-free* if the identity is the only element of finite order; for instance, the groups $\mathbb{Z}^d$, for $d \in \mathbb{N}$, are torsion-free groups.

In the following theorem we characterise the regular linear cellular automata over torsion-free groups.

Theorem 5. *Let G be a torsion-free group and let V be any vector space. A non-zero element $\tau \in \operatorname{LCA}(G; V)$ is regular if and only if it is invertible.*

Proof. It is clear that any invertible element is regular. Let $\tau \in \operatorname{LCA}(G; V) \cong \operatorname{End}(V)[G]$ be non-zero regular. By definition, there exists $\sigma \in \operatorname{LCA}(G; V)$ such that $\tau \sigma \tau = \tau$. As $\operatorname{LCA}(G; V)$ is a ring, the previous is equivalent to

$$\tau(\sigma\tau - 1) = 0,$$

where $1 = 1e$ and $0 = 0e$ are the identity and zero endomorphisms, respectively. Since $\tau \neq 0$, either $\sigma\tau - 1 = 0$, in which case τ is invertible, or $\sigma\tau - 1$ is a zero-divisor. In the latter case, [7, Proposition 6] implies that $\sigma\tau$ has finite order; since G is torsion-free, we must have $\sigma\tau = 1$, so τ is invertible.   □

The *characteristic* of a field $\mathbb{F}$, denoted by char($\mathbb{F}$), is the smallest $k \in \mathbb{N}$ such that

$$\underbrace{1 + 1 + \cdots + 1}_{k \text{ times}} = 0,$$

where 1 is the multiplicative identity of $\mathbb{F}$. If no such k exists we say that $\mathbb{F}$ has characteristic 0.

A group G is *locally finite* if every finitely generated subgroup of G is finite; in particular, the order of every element of G is finite. Examples of such groups are finite groups and infinite direct sums of finite groups.

Theorem 6. *Let G be a group and let V be a finite-dimensional vector space over $\mathbb{F}$. Then, $\mathrm{LCA}(G; V)$ is regular if and only if G is locally finite and char($\mathbb{F}$) $\nmid$ $o(g)$, for all $g \in G$.*

Proof. By [7, Theorem 3] (see also [1, 12]), we have that a group ring $R[G]$ is regular if and only if R is regular, G is locally finite and $o(g)$ is a unit in R for all $g \in G$. In the case of $\mathrm{LCA}(G; V) \cong \mathrm{End}(V)[G]$, since $\dim(V) := n < \infty$, the ring $R := \mathrm{End}(V) \cong M_{n \times n}(\mathbb{F})$ is regular (see [9, Theorem 1.7]. The condition that $o(g)$, seen as the matrix $o(g)I_n$, is a unit in $M_{n \times n}(\mathbb{F})$ is satisfied if and only if $o(g)$ is nonzero in $\mathbb{F}$, which is equivalent to char($\mathbb{F}$) $\nmid o(g)$, for all $g \in G$. $\square$

Corollary 2. *Let G be a group and let V be a finite-dimensional vector space over a field $\mathbb{F}$ of characteristic 0. Then, $\mathrm{LCA}(G; V)$ is regular if and only if G is locally finite.*

Henceforth, we focus on the regular elements of $\mathrm{LCA}(G; V)$ when V is a one-dimensional vector space (i.e. V is just the field $\mathbb{F}$). In this case, $\mathrm{End}_{\mathbb{F}}(\mathbb{F}) \cong \mathbb{F}$, so $\mathrm{LCA}(G; \mathbb{F})$ and $\mathbb{F}[G]$ are isomorphic as $\mathbb{F}$-algebras.

A non-zero element a of a ring R is called *nilpotent* if there exists $n > 0$ such that $a^n = 0$. The following basic result will be quite useful in the rest of this section.

Lemma 2. *Let R be a commutative ring. If $a \in R$ is nilpotent, then a is not a regular element.*

Proof. Let R be a commutative ring and $a \in R$ a nilpotent element. Let $n > 0$ be the smallest integer such that $a^n = 0$. Suppose a is a regular element, so there is $x \in R$ such that $axa = a$. By commutativity, we have $a^2 x = a$. Multiplying both sides of this equation by a^{n-2} we obtain $0 = a^n x = a^{n-1}$, which contradicts the minimality of n. $\square$

Example 6. Suppose that G is a finite abelian group and let $\mathbb{F}$ be a field such that char($\mathbb{F}$) $|$ $|G|$. By Theorem 6, $\mathrm{LCA}(G; \mathbb{F})$ must have elements that are not regular. For example, let $s := \sum_{g \in G} g \in \mathbb{F}[G]$. As $sg = s$, for all $g \in G$, and char($\mathbb{F}$) $|$ $|G|$, we have $s^2 = |G|s = 0$. Clearly, $\mathbb{F}[G]$ is commutative because G is abelian, so, by Lemma 2, s is not a regular element.

We finish this section with the special case when G is the cyclic group $\mathbb{Z}_n$ and $\mathbb{F}$ is a finite field with $\operatorname{char}(\mathbb{F}) \mid n$. By Theorem 6, not all the elements of $\operatorname{LCA}(\mathbb{Z}_n; \mathbb{F})$ are regular, so how many of them are there? In order to count them we need a few technical results about commutative rings.

An *ideal* I of a commutative ring R is a subring such that $rb \in I$ for all $r \in R$, $b \in I$. For any $a \in R$, the *principal ideal* generated by a is the ideal $\langle a \rangle := \{ra : r \in R\}$. A ring is called *local* if it has a unique maximal ideal.

Denote by $\mathbb{F}[x]$ the ring of polynomials with coefficients in $\mathbb{F}$. When $G \cong \mathbb{Z}_n$, we have the following isomorphisms as $\mathbb{F}$-algebras:

$$\operatorname{LCA}(\mathbb{Z}_n; \mathbb{F}) \cong \mathbb{F}[\mathbb{Z}_n] \cong \mathbb{F}[x]/\langle x^n - 1 \rangle,$$

where $\langle x^n - 1 \rangle$ is a principal ideal in $\mathbb{F}[x]$.

Theorem 7. *Let $n \geq 2$ be an integer, and let $\mathbb{F}$ be a finite field of size q such that $\operatorname{char}(\mathbb{F}) \mid n$. Consider the following factorization of $x^n - 1$ into irreducible elements of $\mathbb{F}[x]$:*

$$x^n - 1 = p_1(x)^{m_1} p_2(x)^{m_2} \ldots p_r(x)^{m_r}.$$

For each $i = 1, \ldots, r$, let $d_i := \deg(p_i(x))$. Then, the number of regular elements in $\operatorname{LCA}(\mathbb{Z}_n; \mathbb{F})$ is exactly

$$\prod_{i=1}^{r} \left((q^{d_i} - 1) q^{d_i(m_i - 1)} + 1 \right).$$

Proof. Recall that

$$\operatorname{LCA}(\mathbb{Z}_n; \mathbb{F}) \cong \mathbb{F}[x]/\langle x^n - 1 \rangle.$$

By the Chinese Remainder Theorem,

$$\mathbb{F}[x]/\langle x^n - 1 \rangle \cong \mathbb{F}[x]/\langle p_1(x)^{m_1} \rangle \times \mathbb{F}[x]/\langle p_2(x)^{m_2} \rangle \times \cdots \times \mathbb{F}[x]/\langle p_r(x)^{m_r} \rangle.$$

An element $a = (a_1, \ldots, a_r)$ in the right-hand side of the above isomorphism is a regular element if and only if a_i is a regular element in $\mathbb{F}[x]/\langle p_i(x)^{m_i} \rangle$ for all $i = 1, \ldots, r$.

Fix $m := m_i$, $p(x) = p_i(x)$, and $d := d_i$. Consider the principal ideals $A := \langle p(x) \rangle$ and $B := \langle p(x)^m \rangle$ in $\mathbb{F}[x]$. Then, $\mathbb{F}[x]/B$ is a local ring with unique maximal ideal A/B, and each of its nonzero elements is either nilpotent or a unit (i.e. invertible): in particular, the set of units of $\mathbb{F}[x]/B$ is precisely $(\mathbb{F}[x]/B) - (A/B)$. By the Third Isomorphism Theorem, $(\mathbb{F}[x]/B)/(A/B) \cong (\mathbb{F}[x]/A)$, so

$$|A/B| = \frac{|\mathbb{F}[x]/B|}{|\mathbb{F}[x]/A|} = \frac{q^{dm}}{q^d} = q^{d(m-1)}.$$

Thus, the number of units in $\mathbb{F}[x]/B$ is

$$|(\mathbb{F}[x]/B) - (A/B)| = q^{dm} - q^{d(m-1)} = (q^d - 1) q^{d(m-1)}.$$

As nilpotent elements are not regular by Lemma 2, every regular element of $\mathbb{F}[x]/\langle p_i(x)^{m_i} \rangle$ is zero or a unit. Thus, the number of regular elements in $\mathbb{F}[x]/\langle p_i(x)^{m_i} \rangle$ is $(q^{d_i} - 1) q^{d_i(m_i - 1)} + 1$. $\qquad\square$

Acknowledgments. We thank the referees of this paper for their insightful suggestions and corrections. In particular, we thank the first referee for suggesting the references [1,7,12], which greatly improved the results of Sect. 4.

References

1. Auslander, M.: On regular group rings. Proc. Am. Math. Soc. **8**, 658–664 (1957)
2. Castillo-Ramirez, A., Gadouleau, M.: Ranks of finite semigroups of one-dimensional cellular automata. Semigroup Forum **93**, 347–362 (2016)
3. Castillo-Ramirez, A., Gadouleau, M.: On finite monoids of cellular automata. In: Cook, M., Neary, T. (eds.) AUTOMATA 2016. LNCS, vol. 9664, pp. 90–104. Springer, Cham (2016). doi:10.1007/978-3-319-39300-1_8
4. Castillo-Ramirez, A., Gadouleau, M.: Cellular Automata and Finite Groups. Preprint: arXiv:1610.00532 (2016)
5. Ceccherini-Silberstein, T., Coornaert, M.: Cellular Automata and Groups. Springer Monographs in Mathematics. Springer, Heidelberg (2010)
6. Clifford, A.H., Preston, G.B.: The Algebraic Theory of Semigroups: Mathematical Surveys, vol. 1, no. 7. American Mathematical Society, Providence (1961)
7. Connell, I.G.: On the group ring. Can. J. Math. **15**, 650–685 (1963)
8. Drazin, M.P.: Pseudo-inverses in associative rings and semigroups. Am. Math. Mon. **65**, 506–514 (1958)
9. Goodearl, K.R.: Von Neumann Regular Rings. Monographs and Studies in Mathematics, vol. 4. Pitman Publishing Ltd., London (1979)
10. Howie, J.M.: The subsemigroup generated by the idempotents of a full transformation semigroup. J. Lond. Math. Soc. **s1–41**, 707–716 (1966)
11. Kari, J.: Theory of cellular automata: a survey. Theor. Comput. Sci. **334**, 3–33 (2005)
12. McLaughlin, J.E.: A note on regular group rings. Michigan Math. J. **5**, 127–128 (1958)
13. Zhang, K., Zhang, L.: Generalized reversibility of cellular automata with boundaries. In: Proceedings of the 10th World Congress on Intelligent Control and Automation, Beijing, China (2012)
14. Zhang, K., Zhang, L.: Generalized reversibility of topological dynamical systems and cellular automata. J. Cell. Automata **10**, 425–434 (2015)

Enumerative Results
on the Schröder Pattern Poset

Lapo Cioni and Luca Ferrari[(✉)]

Dipartimento di Matematica e Informatica "U. Dini",
University of Firenze, Firenze, Italy
`lapo.cioni@stud.unifi.it, luca.ferrari@unifi.it`

Abstract. The set of Schröder words (*Schröder language*) is endowed with a natural partial order, which can be conveniently described by interpreting Schröder words as lattice paths. The resulting poset is called the *Schröder pattern poset*. We find closed formulas for the number of Schröder words covering/covered by a given Schröder word in terms of classical parameters of the associated Schröder path. We also enumerate several classes of *Schröder avoiding words* (with respect to the length), i.e. sets of Schröder words which do not contain a given Schröder word.

1 Introduction

In the literature several definitions of patterns in words can be found. In the present article we consider a notion of pattern which is rather natural when words are interpreted as *lattice paths*, by using each letter of the alphabet of the word to encode a possible *step*. The notion of pattern in a lattice path investigated here has been introduced in [1,2], where it has been studied in the case of Dyck paths. Aim of the present work is to find some analogous enumerative results in the case of Schröder paths. In order to make this paper self-contained, we will now briefly recall the main definitions and notations concerning patterns in paths, and we introduce the basic notions concerning the Schröder pattern poset.

For our purposes, a *lattice path* is a path in the discrete plane starting at the origin of a fixed Cartesian coordinate system, ending somewhere on the x-axis, never going below the x-axis and using only a prescribed set of steps Γ. We will refer to such paths as Γ-*paths*. As a word, a Γ-path can be represented by the sequence of the letters encoding the sequence of its steps. In view of this, in the following we will often use the terms "path" and "word" referred to the same object. Classical examples of lattice paths are Dyck, Motzkin and Schröder paths, which are obtained by taking Γ to be the set of steps $\{U, D\}$, $\{U, D, H\}$ and $\{U, D, H_2\}$, respectively (see Fig. 1). Here letters represent the steps $U(p) = (1, 1)$, $D(own) = (1, -1)$, $H(orizontal) = (1, 0)$ and $H_2(orizontal\ of\ length\ 2) = (2, 0)$, respectively.

Partially supported by INdAM-GNCS 2017 project: "Codici di stringhe e matrici non sovrapponibili".

© IFIP International Federation for Information Processing 2017
Published by Springer International Publishing AG 2017. All Rights Reserved
A. Dennunzio et al. (Eds.): AUTOMATA 2017, LNCS 10248, pp. 56–67, 2017.
DOI: 10.1007/978-3-319-58631-1_5

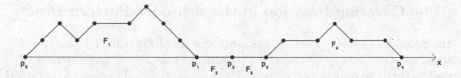

Fig. 1. The Schröder word $UUDUH_2UDDDH_2H_2UH_2UDH_2D$ represented as a Schröder path.

Given a Γ-path P, its *length* is given by the final abscissa of P. Also important is the *word length* of P, which is the length of the word associated with P. For instance, the Schröder path in Fig. 1 has length 22 and word length 17. Notice that the length of a Schröder path is necessarily even; for this reason it is sometimes more meaningful to refer to the *semilength* of a Schröder path.

Given two Γ-paths P and Q, we declare $P \leq Q$ whenever P occurs as a (not necessarily contiguous) subword of Q. In this case, we say that P is a *pattern* of Q. So, for instance, the Schröder path $UH_2UDDH_2UH_2H_2D$ is a pattern of the Schröder path in Fig. 1. When P is not a pattern of Q we will also say that Q *avoids* P. The set $\mathcal{P}_\Gamma$ of all Γ-paths endowed with the above binary relation is clearly a poset.

In the case of Schröder paths, the resulting poset will be denoted $\mathcal{S}$. It is immediate to see that $\mathcal{S}$ has a minimum (the empty path), does not have maximum and is locally finite (i.e. all intervals are finite). Moreover, $\mathcal{S}$ is a ranked poset, and the rank of a Schröder path is given by its semilength. An important fact concerning $\mathcal{S}$ is that it is a *partial well order*, i.e. it contains neither an infinite properly decreasing sequence nor an infinite antichain (this is actually a consequence of a well known theorem by Higman [5]). Notice that this is not the case in another famous pattern poset, the permutation pattern poset, where infinite antichains do exist (see [8]).

The present paper is devoted to the investigation of some structural and enumerative properties of the Schröder pattern poset. Specifically, in Sect. 2 we study the covering relation of $\mathcal{S}$ and in Sect. 3 we enumerate some classes of Schröder paths avoiding a single pattern.

We would like to remark that, even when Γ-paths are interpreted as words over a suitable alphabet, other kinds of *patterns* can be defined which are equally natural and interesting. Just to mention one of the most natural, one can require an occurrence of a pattern to be constituted by *consecutive letters*. This originates what is sometimes called the *factor order*, which has been studied for instance in [4] (in the unrestricted case of all words on a given alphabet). Many papers, such as [3], investigate properties and applications of this more restrictive notion of pattern, also extending it to the case in which the pattern is not required to be a Γ-path itself.

2   The Covering Relation in the Schröder Pattern Poset

In the Schröder pattern poset $\mathcal{S}$, following the usual notation for the covering relation, we write $P \prec Q$ (Q *covers* P) to indicate that $P \leq Q$ and the rank of P is one less than the rank of Q (i.e., $\mathrm{rank}(P) = \mathrm{rank}(Q)-1$). The results contained in the present section concern the enumeration of Schröder paths covered by and covering a given Dyck path Q. We need some notation before stating them.

In a Schröder path Q, let $k+1$ be the number of points of Q (having integer coordinates) lying on the x-axis (call such points $p_0 = (0,0), p_1, \ldots, p_k$). Then Q can be factorized (in a unique way) into k Schröder *factors* $F_1, \ldots, F_k$, each F_i starting at p_{i-1} and ending at p_i. Denote with f_i and h_i the number of U and H_2 steps of factor F_i, respectively. Notice that f_i also equals the number of D steps of the same factor. Let a_i (resp., d_i) be the number of *ascents* (resp. *descents*) in F_i, where an ascent (resp. descent) is a maximal consecutive run of U (resp., D) steps. Moreover, we denote with $p(Q)$ and $v(Q)$ the number of occurrences in Q of a consecutive factor UDU and DUD, respectively. Finally, we denote with $h(Q)$ the total number of *flats* of Q, a flat being a maximal sequence of consecutive H_2 steps. The path depicted in Fig. 1 has 4 factors, and we have $f_1 = 4, f_2 = f_3 = 0, f_4 = 2, h_1 = h_2 = h_3 = 1, h_4 = 2, a_1 = 3, a_2 = a_3 = 0, a_4 = 2, d_1 = 2, d_2 = d_3 = 0, d_4 = 2, p(Q) = 1, v(Q) = 0$ and $h(Q) = 4$.

Proposition 1. *If $Q = F_1 F_2 \cdots F_k$ is a Schröder path with k factors, with F_i having a_i ascents and d_i descents, then the number of Schröder paths covered by Q is given by*

$$\sum_{1 \leq i \leq j \leq k} d_i a_j - p(Q) - v(Q) + h(Q). \tag{1}$$

Proof. There are two (mutually exclusive) ways to obtain a Schröder path covered by Q, namely:

1. by removing a H_2 step, or
2. by removing a U step and a D step.

   We examine the two cases separately.

1. It is immediate to observe that one obtains the same path by removing any of the steps belonging to the same flat, whereas removing a step from different flats gives rise to different paths. Therefore, the number of distinct Schröder paths obtained from Q by removing an H_2 step is $h(Q)$.
2. We wish to prove that there are $\sum_{1 \leq i \leq j \leq k} d_i a_j - p(Q) - v(Q)$ ways to remove a U step and a D step from Q and to obtain another Schröder path. We will proceed by induction on the number k of factors of Q. If $k = 1$, then necessarily Q starts with a U step and ends with a D step (otherwise $Q = H_2$, which has no U and D steps). Observe that, in this case, we can remove any of the U steps and any of the D steps and the resulting path is still a Schröder path. Removing steps from the same ascent (and from the same descent) returns the same path, so we have a_1 possible choices to remove a U step

and d_1 possible choices to remove a D step. However, there are some special cases in which, though removing from different ascents or descents, we obtain the same path. Specifically, if we have a consecutive string UDU in Q, then removing from Q the UD of such a string returns the same path as removing the DU, in spite of the fact that the two U steps belong to different ascents. In a similar way, the presence of a factor DUD in Q gives the possibility of getting the same Schröder path by removing steps from different descents. To avoid overcount, we thus have to subtract the number of consecutive strings UDU and DUD of Q, thus obtaining a total of $d_1a_1 - p(Q) - v(Q)$ paths. Now suppose that $k > 1$. There are three distinct cases to analyze.

- If we remove both the U and the D steps from the prefix $F_1 \cdots F_{k-1}$ of Q consisting of the first $k - 1$ factors (which is a Schröder path in itself, of course), by induction we have $\sum_{1 \le i \le j \le k-1} d_i a_j - p(F_1 \cdots F_{k-1}) - v(F_1 \cdots F_{k-1})$ distinct choices.
- If we remove both the U and the D steps from the last factor F_k, using the same argument as in the case $k = 1$ we get $d_k a_k - p(F_k) - v(F_k)$ distinct paths.
- Finally, suppose we choose to remove the D step from the prefix $F_1 \cdots F_{k-1}$ and the U step from the last factor F_k (notice that we are not allowed to do the opposite, otherwise the resulting path would not be Schröder). In this case we have a_k possible choices for U and $\sum_{i=1}^{k-1} d_i$ possible choices for D. Once again, however, there are some paths that are overcounted, occurring when $F_1 \cdots F_{k-1}$ and F_k share a consecutive UDU or a consecutive DUD. A quick look shows that this overcount is corrected by subtracting the number of such shared occurrences of consecutive UDU and DUD.

The sum of the above three cases gives the required expression.

Finally, summing up the quantities in the two cases above, we obtain precisely formula (1). ∎

Remark. If Q is a Dyck path, then $h(Q) = 0$, and formula (1) reduces to the analogous formula for Dyck paths obtained in [1,2], since a Schröder path covered by a Dyck path is necessarily a Dyck path.

Proposition 2. *Let $P = F_1 \cdots F_k$ be a Schröder path having k factors. Denote with f_i the number of U steps in the factor F_i (this is also the number of D steps in F_i) and with h_i the number of H_2 steps in F_i. Moreover, let ℓ be the word length of P. Then the number of Schröder paths covering P is given by*

$$2 + \ell + \sum_{\substack{(i,j) \\ 1 \le i \le j \le k}} (f_i + h_i)(f_j + h_j). \tag{2}$$

Proof. We have two options to obtain a Schröder path covering P:

1. either we add a H_2 step, or
2. we add a U step and a D step.

As in the previous proposition, we examine the two cases separately.

1. Adding a new H_2 step in any point of a flat of P returns the same path. Hence, in order to obtain distinct paths, we can add a H_2 step either before a U step, or before a D step, or at the end of P. Thus we have a total of

$$2 \sum_{i=1}^{k} f_i + 1$$

paths covering P in this case.

2. We start by observing that adding a U step in any point of an ascent returns the same path, and the same holds for D steps (with ascents replaced by descents). Suppose to add a new U step to P first. In order to obtain distinct paths, we can add U either before a D step, or before a H_2 step, or at the end of P.

   If a U step is added before a D step in F_i, we observe that we cannot add the new D step in a factor F_j, with $j < i$, otherwise the path would fall below the x-axis. With this constraint in mind, we are now allowed to add the new D step either before a U or before a H_2 or at the end of the path. However, in the first of the three previous cases, we cannot of course add the new D step before the first allowed U (i.e., at the beginning of the factor); moreover, adding the new D right before the new U step just added would produce a substring DUD, which can be obtained also by first adding the U step in the following available position of P and then adding the D step immediately after it. Thus, in this case, the number of paths covering P is obtained by considering the number of possible choices for U to be added in F_i, which is f_i, and the number of possible choices for D, which is $\sum_{j \geq i}(f_j + h_j)$, and so it is

$$\sum_{\substack{(i,j) \\ 1 \leq i \leq j \leq k}} f_i(f_j + h_j).$$

   If a U step is added before a H_2 step in F_i, as in the previous case, we cannot add the new D step in a factor F_j, with $j < i$. We can now add the new D step either before a U (except for the first U of F_i, of course), or before a H_2 or at the end of P. So, in this case, the number of paths covering P is given by

$$\sum_{i=1}^{k} h_i \cdot \left(1 + \sum_{j=i}^{k}(f_j + h_j) \right).$$

   Finally, if we add the new U step at the end of P, then the new D step must necessarily be added after it.

Summing up, we therefore obtain the following expression for the total number of paths covering P:

$$2\sum_{i=1}^{k} f_i + 1 + \sum_{\substack{(i,j) \\ 1 \le i \le j \le k}} f_i(f_j + h_j) + \sum_{i=1}^{k} h_i \cdot \left(1 + \sum_{j=i}^{k}(f_j + h_j)\right) + 1$$

$$= 2 + \sum_{i=1}^{k}(2f_i + h_i) + \sum_{\substack{(i,j) \\ 1 \le i \le j \le k}} f_i(f_j + h_j) + \sum_{\substack{(i,j) \\ 1 \le i \le j \le k}} h_i(f_j + h_j)$$

$$= 2 + \ell + \sum_{\substack{(i,j) \\ 1 \le i \le j \le k}} (f_i + h_i)(f_j + h_j), \tag{3}$$

which is formula (2).   ∎

Remark 1. Notice that $f_i + h_i$ is the semilength of the factor F_i. Denoting it with φ_i, formula (2) can be equivalently written as

$$2 + \ell + \sum_{\substack{(i,j) \\ 1 \le i \le j \le k}} \varphi_i \varphi_j.$$

Remark 2. If P is a Dyck path, then, in the first expression in the chain of equalities (3), the summand $2\sum_{i=1}^{k} f_i + 1$ gives the number of non-Dyck paths covering P (i.e., those having one H_2 step), and the remaining summands give the number of Dyck paths covering P. Also in this case, recalling that $h_i = 0$ for all i, we recover the analogous formula for Dyck paths obtained in [1,2].

3   Enumerative Results on Pattern Avoiding Schröder Paths

Main goal of the present section is to enumerate several classes of Schröder paths avoiding a given pattern. For any Schröder path P, denote with $\mathcal{S}_n(P)$ the set of Schröder paths of semilength n avoiding P, and let $s_n(P) = |\mathcal{S}_n(P)|$ be its cardinality. It is completely trivial to observe that

- $s_n(\emptyset) = 0$;
- $s_n(H_2) = C_n$, where $C_n = \frac{1}{n+1}\binom{2n}{n}$ is the n-th Catalan number (sequence A000108 of [7]), counting the number of Dyck paths of semilength n;
- $s_n(UD) = 1$ when $n > 0$.

Starting from patterns of semilength 2, we get some interesting enumerative results. In the next subsections we define several classes of Schröder paths avoiding a single pattern, each of which suitably generalizes the case of a pattern of semilength 2. In all cases, after having described a general enumeration formula, we illustrate it in the specific case of the relevant pattern of semilength 2.

Before delving into computations we state an important lemma, which in several cases reduces the enumeration of pattern avoiding Schröder paths to the case of pattern avoiding Dyck paths. In this lemma, as well as in several subsequent proofs, we will deal with the *multiset coefficient* $\left(\binom{n}{k}\right)$, counting the number of multisets of cardinality k of a set of cardinality n. As it is well known, the multiset coefficients can be expressed in terms of the binomial coefficients, namely $\left(\binom{n}{k}\right) = \binom{n+k-1}{k}$.

Lemma 1. *Given a Dyck path P, denote with $d_n(P)$ the number of Dyck paths of semilength n avoiding P. Then*

$$s_n(P) = \sum_{h=0}^{n} \binom{n+h}{n-h} d_h(P). \tag{4}$$

Proof. Let Q be a Schröder path. Clearly Q avoids P if and only if the Dyck path $\tilde{Q}$ obtained from Q by deleting all horizontal steps avoids P. Therefore, the set of Schröder paths of semilength n with $n - h$ horizontal steps avoiding P is obtained by taking in all possible way a Dyck path of semilength h avoiding P and then adding to it $n - h$ horizontal steps in all possible ways. Observe that, in a Dyck path of semilength h, one has $2h + 1$ possible sites where to insert a horizontal step, and any number of horizontal steps can be inserted into the same site. Thus, if $n - h$ horizontal steps have to be inserted, it is necessary to select a multiset of cardinality $n - h$ from the set of the possible $2h + 1$ sites. This can be done in $\binom{(2h+1)+(n-h)-1}{n-h} = \binom{n+h}{n-h}$ ways, as it is well known. Since h can be chosen arbitrarily in the set $\{0, 1, 2, \ldots, n\}$, the total number of Schröder paths of semilength n avoiding P is given by formula (4), as desired. ∎

3.1 The Pattern $(UD)^k$

Since $(UD)^k$ is a Dyck path, this case can be seen as an immediate consequence of Lemma 1 together with the results of [1,2].

Proposition 3. *For $i, j \geq 1$, let $N_{i,j} = \frac{1}{i}\binom{i}{j}\binom{i}{j-1}$ be the Narayana numbers (sequence A001263 of [7]). Extend such a sequence by setting $N_{0,0} = 1$ and $N_{i,0} = N_{0,j} = 0$, for all $i, j > 0$. Then*

$$s_n((UD)^k) = \sum_{h=0}^{n} \sum_{j=0}^{k-1} \binom{n+h}{n-h} N_{h,j}. \tag{5}$$

The case $k = 2$ gives rise to an interesting situation. In fact, for the pattern $UDUD$, recalling that $N_{0,0} = 1$, $N_{i,0} = 0$ and $N_{i,1} = 1$ for all $i > 0$, formula (5) gives

$$s_n(UDUD) = \sum_{h=0}^{n} \binom{n+h}{n-h} N_{h,0} + \sum_{h=0}^{n} \binom{n+h}{n-h} N_{h,1}$$

$$= 1 + \sum_{h=1}^{n} \binom{n+h}{n-h} = \sum_{h=0}^{n} \binom{2n-h}{h}. \tag{6}$$

Since it is well known that Fibonacci numbers $(F_n)_n$ (sequence A000045 in [7]) can be expressed in terms of binomial coefficients as[1] $F_{n+1} = \sum_{k \geq 0} \binom{n-k}{k}$, we get $s_n(UDUD) = F_{2n+1}$, i.e. Schröder paths avoiding $UDUD$ are counted by Fibonacci numbers having odd index (sequence A122367 of [7]).

Remark. Notice that, for a Schröder path, avoiding the (Schröder) path $UDUD$ is equivalent to avoiding the (non-Schröder) path DU. As suggested by one of the referees, a simple combinatorial argument to count Schröder words of semilength n avoiding the subword DU is the following: if the word contains k H_2 steps, then it can be constructed by taking the word $U^{n-k}D^{n-k}$ and inserting k H_2 steps. Taking into account all possibilities, and summing over k, gives precisely formula (6).

3.2   The Pattern $U^k D^k$

This case is similar to the previous one, in that it can be easily inferred from Lemma 1, since the generic pattern of the class is a Dyck path. Thus, applying the above mentioned lemma and using results of [1,2], we obtain the following result.

Proposition 4. *For all $k \geq 0$, we have*

$$s_n(U^k D^k) = \sum_{h=0}^{k-1} \binom{n+h}{n-h} C_h + \binom{n+k}{n-k}(C_k - 1)$$

$$+ \sum_{h=k+1}^{\min\{2k-1,n\}} \sum_{j \geq 1} \binom{n+h}{n-h} b_{k-j,h-k+j}^2, \qquad (7)$$

where the C_n's are the Catalan numbers and the $b_{i,j}$'s are the ballot numbers (sequence A009766 of [7]).

Setting $k = 2$ in formula (7), for Schröder paths of semilength $n \geq 3$ avoiding $UUDD$ we obtain the following polynomial of degree 4 (sequence A027927 of [7]):

$$s_n(UUDD) = 1 + \binom{n+1}{n-1} + \binom{n+2}{n-2} = 1 + \frac{n(n+1)(n^2 + n + 10)}{24}.$$

3.3   The Pattern H_2^k

In this case the generic pattern of this class is not a Dyck path. However, we are able to give a direct argument to count Schröder paths avoiding H_2^k.

---

[1] Notice that the sum contains only a finite number of nonzero terms, since the binomial coefficients vanish when $k > \lfloor n/2 \rfloor$.

Proposition 5. *For all $n, k \geq 0$, we have*

$$s_n(H_2^k) = \sum_{i=0}^{k-1} \binom{2n-i}{i} C_{n-i}. \tag{8}$$

Proof. We observe that a Schröder path avoids the pattern H_2^k if and only if it has at most $k - 1$ H_2 steps. Thus, the set of all Schröder paths of semilength n avoiding H_2^k can be obtained by taking the set of all Dyck paths of semilength $n - i$ and inserting in all possible ways i H_2 steps, for i running from 0 to $k - 1$. Since in a Dyck path of semilength $n - i$ there are precisely $2n - 2i + 1$ points in which inserting the horizontal steps, and we have to insert i horizontal steps (possibly inserting more than one H_2 step in the same place), we get

$$s_n(H_2^k) = \sum_{i=0}^{k-1} \left(\left(\binom{2n-2i+1}{i} \right) \right) C_{n-i} = \sum_{i=0}^{k-1} \binom{2n-2i+1+i-1}{i} C_{n-i}$$

$$= \sum_{i=0}^{k-1} \binom{2n-i}{i} C_{n-i},$$

as desired.   ∎

When $k = 2$ we obtain the following special case:

$$s_n(H_2 H_2) = C_n + (2n-1)C_{n-1} = \frac{n+3}{2} C_n = \frac{n+3}{2(n+1)} \binom{2n}{n},$$

which is valid for $n \geq 1$. This is sequence A189176 of [7], whose generating function is $\frac{1-5x+4x^2-(1-5x)\sqrt{1-4x}}{2x(1-4x)}$, and can be also obtained as the row sums of a certain Riordan matrix (see [7] for details).

3.4   The Pattern $UH_2^{k-1}D$

This class of patterns requires a little bit more care, nevertheless we are able to get a rather neat enumeration formula.

Proposition 6. *For all $n \geq 0, k > 1$, we have*

$$s_n(UH_2^{k-1}D) = 1 + \sum_{h=1}^{n} \sum_{i=0}^{\min\{k-2,n-h\}} \left(\left(\binom{2h-1}{i} \right) \right) (n-h-i+1)C_h. \tag{9}$$

Proof. Let P be a Schröder path of semilength n avoiding $UH_2^{k-1}D$. If P does not contain U steps, then necessarily $P = H_2^n$. Otherwise, P can be decomposed into three subpaths, namely:

- a prefix, consisting of a (possibly empty) sequence of horizontal steps;
- a path starting with the first U step and ending with the last D step (necessarily not empty);
- a suffix, consisting of a (possibly empty) sequence of horizontal steps.

The central portion of P in the above decomposition has at most $k - 2$ horizontal steps. Thus it can be obtained starting from a Dyck path of semilength h, for some $1 \leq h \leq n$, and adding i horizontal steps, for some $0 \leq i \leq \min\{k - 2, n - h\}$. Such horizontal steps can be inserted into $2h - 1$ possible sites (here we have to exclude the starting and the ending points, since the subpath is required to start with a U and end with a D), with the possibility of inserting several steps into the same site, as usual. The resulting path has therefore semilength $h + i$, and has to be completed by adding a suitable number of horizontal steps at the beginning and at the end, to obtain a Schröder path of semilength n: there are $n - h - i + 1$ possible ways to do it. We thus obtain formula (9) for $s_n(UH_2^{k-1}D)$, as desired.   ∎

Formula (9) becomes much simpler in the special case $k = 2$ of Schröder paths of semilength n avoiding UH_2D. Indeed we obtain:

$$s_n(UH_2D) = 1 + \sum_{h=1}^{n}(n - h + 1)C_h = \sum_{h=0}^{n}C_h(n - h) + \sum_{h=0}^{n}C_h - n.$$

In this case, we can find an interesting expression for the generating function of these coefficients in terms of the generating function $C(x) = \sum_{n \geq 0}C_nx^n$ of Catalan numbers, which provides an easy way to compute them:

$$\sum_{n \geq 0}s_n(UH_2D)x^n = C(x) \cdot \sum_{n \geq 0}nx^n + C(x) \cdot \sum_{n \geq 0}x^n - \sum_{n \geq 0}nx^n$$

$$= C(x)\left(\frac{x}{(1 - x)^2} + \frac{1}{1 - x}\right) - \frac{x}{(1 - x)^2}$$

$$= \frac{1}{(1 - x)^2}(C(x) - x).$$

Roughly speaking, the above generating function tells us that $s_n(UH_2D)$ is given by the partial sums of the partial sums of the sequence of Catalan numbers where C_1 is replaced by 0. The associated number sequence starts with $1, 2, 5, 13, 35, 99, 295, \ldots$ and does not appear in [7].

3.5   The Pattern $H_2^{k-1}UD$

The last class of patterns we consider is the most challenging one. It gives rise to an enumeration formula which is certainly less appealing than the previous ones. Due to space limitation, we will just sketch its proof and simply state the special case corresponding to $k = 2$. Before illustrating our final results, we need to introduce a couple of notations.

We denote with $P_{k,h}$ the number of Dyck prefixes of length k ending at height h. Notice that we can express these coefficients in terms of the ballot numbers $b_{i,j} = \frac{i-j+1}{i+1}\binom{i+j}{i}$, counting the number of Dyck prefixes with i up steps and j down steps, as follows:

$$P_{h,k} = \sum_{\substack{i,j \\ i-j=h \\ i+j=k}} b_{i,j}.$$

Moreover, we denote with $S_{n,q}$ the number of Schröder paths of semilength n having exactly q horizontal steps. Since each such path can be uniquely determined by a Dyck path of semilength $n-q$ with q horizontal steps added, we have a rather easy way to compute $S_{n,q}$ in terms of the Catalan numbers C_n:

$$S_{n,q} = C_{n-q}\left(\binom{2(n-q)+1}{q}\right) = \binom{2n-q}{q}C_{n-q}.$$

Proposition 7. *For all $n \geq 0, k > 1$, we have*

$$s_n(H_2^{k-1}UD) = \sum_{q=0}^{k-2} S_{n,q}$$

$$+ \sum_{p=0}^{2n-2k+2}\sum_{h=0}^{2n-p-2k+2} P_{p,h}\left(\binom{p+1}{k-2}\right)\left(\binom{\frac{2n-h-p-2k+4}{2}}{h}\right). \quad (10)$$

Proof. Let P be a Schröder path of semilength n. If P contains less than $k-1$ horizontal steps, then it necessarily avoids $H_2^{k-1}UD$. This gives the term $\sum_{q=0}^{k-2} S_{n,q}$ in the r.h.s. of (10). If instead P contains at least $k-1$ horizontal steps, then, in order to avoid $H_2^{k-1}UD$, it has to be decomposable into a Schröder prefix ending at some height h and having exactly $k-2$ horizontal steps, and a suffix starting with a horizontal step followed exclusively by H_2 and D step (with exactly h D steps). The generic Schröder prefix of the required form can be obtained by taking a Dyck prefix of length p, for some $0 \leq p \leq 2n-2k+2$, and adding $k-2$ H_2 steps in all possible ways. We thus get a total of $P_{p,h} \cdot \left(\binom{p+1}{k-2}\right)$ Schröder prefixes of length $p+2k-4$ ending at height h and containing exactly $k-2$ H_2 steps. The generic suffix of the required form contains h D steps and $\frac{2n-h-p-2k+4}{2}$ H_2 steps. Such a suffix can be obtained by inserting the h down steps into the sequence of horizontal steps in all possible ways. Since the first step has to be a horizontal one, this gives a total of $\left(\binom{\frac{2n-h-p-2k+4}{2}}{h}\right)$ allowed suffixes. Putting together all the contributions, we get the desired expression for $s_n(H_2^{k-1}UD)$. ∎

Specializing to $k = 2$ we obtain

$$s_n(H_2UD) = C_n + \sum_{p=0}^{2n-2}\sum_{h=0}^{2n-p-2} P_{p,h}\left(\binom{\frac{2n-h-p}{2}}{h}\right).$$

4 Suggestions for Further Work

It would be very interesting to investigate in more detail the structural properties of the Schröder pattern poset. A typical question in this context concerns the computation of the Möbius function, which is still open even in the Dyck pattern poset. Another (partially related) issue is the enumeration of (saturated) chains. More generally, can we say anything about the order structure of intervals (for instance, is it possible to determine when they are lattices?)?

Concerning the enumeration of pattern avoiding classes, the next step would be to count classes of Schröder words simultaneously avoiding two or more patterns.

Finally, it would be nice to have information on the asymptotic behavior of integer sequences counting pattern avoiding Schröder words. In the Dyck case, all the sequences which count Dyck words avoiding a single pattern P have the same asymptotic behavior (which is roughly exponential in the length of P). This is in contrast, for instance with the permutation pattern poset, where the asymptotic behavior of a class of pattern avoiding permutations depends on the patterns to be avoided (this is the ex Stanley-Wilf conjecture, proven by Marcus and Tardos [6]). What does it happen in the Schröder pattern poset?

References

1. Bacher, A., Bernini, A., Ferrari, L., Gunby, B., Pinzani, R., West, J.: The Dyck pattern poset. Discrete Math. **321**, 12–23 (2014)
2. Bernini, A., Ferrari, L., Pinzani, R., West, J.: Pattern avoiding Dyck paths. Discrete Math. Theoret. Comput. Sci. Proc. **AS**, 683–694 (2013)
3. Bilotta, S., Grazzini, E., Pergola, E., Pinzani, R.: Avoiding cross-bifix-free binary words. Acta Inform. **50**, 157–173 (2013)
4. Björner, A.: The Möbius function of factor order. Theoret. Comput. Sci. **117**, 91–98 (1993)
5. Higman, G.: Ordering by divisibility in abstract algebras. Proc. London Math. Soc. **3**, 326–336 (1952)
6. Marcus, A., Tardos, G.: Excluded permutation matrices and the Stanley-Wilf conjecture. J. Combin. Theory Ser. A **107**, 153–160 (2004)
7. N. J. A. Sloane, The On-Line Encyclopedia of Integer Sequences, electronically available at oeis.org
8. Spielman, D.A., Bóna, M.: An infinite antichain of permutations. Electron. J. Combin. **7**, 4 (2000). #N2

Canonical Form of Gray Codes in N-cubes

Sylvain Contassot-Vivier[1][(✉)] and Jean-François Couchot[2]

[1] Université de Lorraine, LORIA, UMR 7503, 54506 Vandoeuvre-lès-Nancy, France
Sylvain.Contassotvivier@loria.fr
[2] FEMTO-ST Institute, Univ. Bourgogne Franche-Comté, Besançon, France

Abstract. In previous works, the idea of walking into a N-cube where a balanced Hamiltonian cycle have been removed has been proposed as the basis of a chaotic PRNG whose chaotic behavior has been proven. However, the construction and selection of the most suited balanced Hamiltonian cycles implies practical and theoretical issues. We propose in this paper a canonical form for representing isomorphic Gray codes. It provides a drastic complexity reduction of the exploration of all the Hamiltonian cycles and we discuss some criteria for the selection of the most suited cycles for use in our chaotic PRNG.

1 Introduction

The problem of designing Pseudo-Random Number Generators (PRNG) that satisfy the probabilistic properties to produce a uniform distribution is difficult. Moreover, the knowledge of the generation algorithm and any sequence of previously generated bits should not constitute a sufficient piece of information to predict the next generated bits without knowing initial conditions. In order to build such PRNG, some studies have focused on the use of chaotic systems [2,6,7].

In a previous work [4], some of the authors have proposed a PRNG based on random walk in a N-cube where a balanced Hamiltonian cycle has been removed, and its chaotic nature has been proved. Moreover, it has been shown that the removed Hamiltonian cycle should be balanced in order to produce more efficient PRNG. Balanced Hamiltonian cycles are cycles in which the numbers of occurrences of the traversed dimensions are equal or differ at most by 2. In [8], the authors have proposed an approach that provides a subset of all the Hamiltonian cycles. This approach is however undeterministic and the cardinal number of the produced subset is dramatically small compared to the one of all the Hamiltonian cycles. In some sense, it is a partial solution of finding Hamiltonian cycles.

The undeterministic aspect of this approach has been tackled in [3] where we have proposed a particularization of it. This new procedure succeed to find balanced Hamiltonian cycles for any dimension N and solves this issue. Nevertheless, pursuing our objective to enhance the specification of the Hamiltonian cycles most suited to the use in our PRNG, we have been confronted to the fact

A. Dennunzio et al. (Eds.): AUTOMATA 2017, LNCS 10248, pp. 68–80, 2017.
DOI: 10.1007/978-3-319-58631-1_6

that procedure detailed in [3] cannot produce all non-isomorphic balanced codes: it is indeed a particularization of a partial solution.

In [10], the author proposed an approach to produce all the cycles of a graph. This work may thus solve the problem of finding a large set of balanced Hamiltonian given a dimension N. However, the approach suffers from being too exhaustive and cannot be applied as soon as the dimension of the N-cube is larger than 5. One solution could be to study cycles, whose embedding into PRNG gives distinct behaviors, *i.e.* which do not belong in the same class w.r.t an equivalence relation. For that, this work proposes a canonical form dedicated to cycles and its application to the generation of a large set of balanced Hamiltonian cycles.

This paper presents these two elements. In the following section is presented the canonical form of Hamiltonian cycles, followed in Sect. 3 by the description of our novel algorithm. The practical interest of the algorithm is discussed in Sect. 4.

2   Canonical Form of Gray Codes

Let $S_N = \{1, ..., N\}^{2^N}$, the set of sequences of length 2^N with values in $\{1, ..., N\}$. Let $H_N \subset S_N$, the set of sequences describing Hamiltonian cycles in a N-cube. Each of these sequences gives the succession of the dimensions followed by the path. Any Hamiltonian cycle of H_N can be written as $h = (h_1, ..., h_{2^N})$. Also, we remind the reader that a Hamiltonian cycle in a N-cube is a Gray code.

We call the *canonical form* of a Hamiltonian cycle, an equivalent description of the cycle that is obtained, through a specific computation process, for all its isomorphic cycles.

Before describing our computation process of the canonical forms, we provide below an overview of the different cases of isomorphism between cycles.

2.1   Isomorphic Cycles

Intuitively, Hamiltonian cycles are isomorphic to each other when the paths they describe can be topologically superposed. Indeed, a same Hamiltonian cycle can be expressed in many sequences according to some simple (global) transformations of the N-cube, leading to a set of isomorphic cycles. We list below the different transformations that can be applied to a sequence to produce isomorphic cycles.

First of all, it can be noticed that describing a cycle by the sequence of the traversed dimensions in the N-cube does not specify any starting vertex. So, a sequence does not represent only a single cycle but the 2^N cycles that are isomorphic up to the starting position in the N-cube.

In a similar way, applying a cyclic shift to a sequence, in any direction, is equivalent to change only its starting vertex, but this does not change the path topology. So, shifted sequences are also isomorphic cycles.

Moreover, as the N-cubes considered in the scope of this paper are not oriented, the direction of the cycle is not significant and then, an isomorphic cycle is obtained by inverting the order of a sequence.

Finally, cycles can also be isomorphic up to rotations/symmetries, which are obtained by renumbering the dimensions of the N-cube. For example, exchanging dimensions 2 and 3 in a 3-cube is similar to performing a 90° rotation around dimension 1. In the following, that operation may also be referred to as the relabeling of a sequence since it only changes the dimensions labels. It is worth noticing that some dimensions relabelings are equivalent to the sequence inversion combined with a cyclic shift.

In order to define the canonical form of Hamiltonian cycle, we need to introduce some functions over H_N.

2.2 Preliminary Tools

Let $R : H_N \rightarrow H_N$, the function that renumbers a Hamiltonian cycle h to a sequence $R(h)$ by mapping the successive distinct values (dimensions) of h to the ordered values from 1 to N. So, the first value h_1 of h is necessarily mapped to 1, then the first distinct value in the remaining of h (that is $(h_2, \ldots, h_{2^N})$) is mapped to 2, and so on. As function R applies a renumbering, it follows that $\forall i, j \in \{1, \ldots, 2^N\}, h_i = h_j \Leftrightarrow R(h)_i = R(h)_j$.

The effect of function R is to apply rotations/symmetries to a sequence, by relabeling the dimensions of the N-cube, in order to express it in a specific order of the traversed dimensions, without modifying topology of the path. So, this function is an automorphism on H_N.

As an example, if we have N = 3 and the sequence $h = (2, 3, 1, 3, 2, 3, 1, 3)$, then $R(h) = (1, 2, 3, 2, 1, 2, 3, 2)$. So, the dimensions 1, 2 and 3 are respectively replaced by (relabeled) 3, 1 and 2 (as shown in Fig. 1), where the three dimensions labels and the starting vertex are fixed. It can be seen that both sequences are isomorphic up to a rotation around dimension 1 and an orientation inversion.

As the lexicographic order over sequences of length N provides a total order on H_N, the results of R are totally ordered. So, for any subset X of H_N there

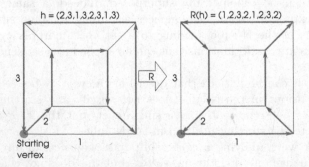

Fig. 1. Application of function R on sequence $h = (2, 3, 1, 3, 2, 3, 1, 3)$.

exists a unique minimal value of the results of R applied to any $h \in X$. This property is used in the computation of our canonical form.

Let $D : H_N \times \{1, \ldots, 2^N\} \rightarrow H_N$, the function that associates to a sequence $h = (h_1, \ldots, h_{2^N})$ and an integer k, the sequence $D(h, k) = (h_k, h_{k+1}, \ldots, h_{2^N}, h_1, h_2, \ldots, h_{k-1})$, which is h after $k - 1$ successive left cyclic shifts, so that h_k becomes the first value of the sequence.

The effect of function D is simply to change the starting point of the sequence, without modifying the cycle itself, as can be seen on Fig. 2. As well as function R, this function is also an automorphism on H_N and it is also used to compute our canonical form of isomorphic cycles.

Going back to our previous example sequence $h = (2, 3, 1, 3, 2, 3, 1, 3)$ and choosing $k = 3$, we obtain $D(h, 3) = (1, 3, 2, 3, 1, 3, 2, 3)$.

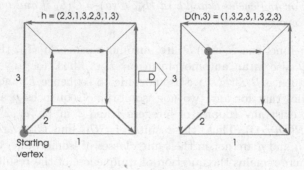

Fig. 2. Application of function D on sequence $h = (2, 3, 1, 3, 2, 3, 1, 3)$ and $k = 3$.

Let $W : H_n \times \{1, \ldots, 2^N\} \rightarrow \{1, \ldots, 2^N\}$, the function that associates to a sequence h and an integer k, the length of the minimal sub-sequence of cycle h starting at h_k and containing all the values in $\{1, \ldots, n\}$.

Getting back to our example h and choosing $k = 4$, we have $W((2, 3, 1, 3, 2, 3, 1, 3), 4) = 4$ as the smallest sub-sequence containing $\{1, 2, 3\}$ starting from $h_4 = 3$ in h is $(2, 3, 1, \underline{3, 2, 3, 1}, 3)$, that is to say $(3, 2, 3, 1)$, whose length is 4. In the same way, we have $W((2, 3, 1, 3, 2, 3, 1, 3), 7) = 3$ as the minimal sub-sequence from $h_7 = 1$ is $(1, 3, 2)$. However, we can notice too that $W((2, 3, 1, 3, 2, 3, 1, 3), 8) = 4$.

In [1], Bykov uses this notion of minimal sub-sequence containing all the dimensions of the N-cube to define the *window width* of a sequence h. It corresponds to the maximal value of function W over all the possible starting points in the sequence. It provides information about the *local balance* between the traversed dimensions along the cycle. This window width can be defined by $M(h)$, for $h \in H_N$ as:

$$M(h) = \max_{k \in \{1, \ldots, 2^N\}} W(h, k) \tag{1}$$

2.3 Canonical Form

The function $C : H_N \rightarrow H_N$, defined by:

$$C(h) = \min \{R(D(h, k)) | \, k \text{ is s.t. } W(h, k) = M(h)\} \tag{2}$$

produces the canonical form of any sequence from H_N. Notice that this set is ordered according to the aforementioned lexical ordering, which is total.

The role of the C function is to provide a unique representative of for each class of Hamiltonian cycle. By class of Hamiltonian cycle, we mean the set of isomorphic Hamiltonian cycles according to translations (changing the starting point of the sequence) and rotations/symmetries (changing the dimensions labels). So, we have the following theorem.

Theorem 1. *For any cycles a and b in H_N, $C(a) = C(b)$ if and only if a and b are isomorphic cycles.*

Proof. As both functions R and D are automorphisms on H_N, the composite function $R \circ D$ also is an automorphism on H_N. Thus, for any integer $k \in \{1, \ldots, 2^N\}$, sequence $R(D(h, k))$ is isomorphic to sequence h, and so is $C(h)$. Also, this implies that for any two non-isomorphic sequences h and g in H_N, there does not exist any couples of integers i and j in $\{1, \ldots, 2^N\}$ such that $R(D(h, i)) = R(D(g, j))$. Thus, the results of $C(h)$ and $C(g)$ are necessarily different when h and g are not in the same classes of isomorphic cycles.

However, there remains the question of uniqueness of the result of C for all sequences in a same class of H_N. That property induces that for any two isomorphic sequences h and g in H_N, there exist two integers i and j in $\{1, \ldots, 2^N\}$ such that $R(D(h, i)) = R(D(g, j))$. From the previous observations, it is obvious that $R(D(h, k))$ is isomorphic to $R(D(g, j))$, but we have to show that for some adequately chosen i and j, they are *identical* sequences.

As a first step, let us consider two sequences $h = (h_1, \ldots, h_l)$ and $g = (g_1, \ldots, g_l)$ that are isomorphic only up to rotations/symmetries. As such transformations can be expressed by dimensions relabeling, it follows that g and h are mutual relabelings of each others:

$$h_1 \leftrightarrow g_1, \quad h_2 \leftrightarrow g_2, \quad \ldots, \quad h_l \leftrightarrow g_l \tag{3}$$

and

$$\forall i, j \in \{1, \ldots, l\} \quad h_i = h_j \Leftrightarrow g_i = g_j \tag{4}$$

Moreover, $R(h)$ and $R(g)$ are also respective relabelings of h and g. The fact that $R(h) = R(g)$ is ensured by the ordered relabeling over $\{1, \ldots, n\}$. Indeed, as the relabeling follows the numerical order of integers, it produces the same sequence for h and g according to the total lexicographic order over H_N:

$$\begin{aligned} h_1 \rightarrow 1, g_1 \rightarrow 1 \\ h_2 \rightarrow 2, g_2 \rightarrow 2 \end{aligned} \tag{5}$$

and due to (4), we have:

$$\forall i \in \{3, \ldots, l\}, k \in \{1, \ldots, n\}, \quad h_i \to k \Leftrightarrow g_i \to k \tag{6}$$

Thus, function R produces the same result for sequences that are isomorphic up to rotations/symmetries.

The next step consists in taking into account cyclic shifts between sequences. Solving this problem is similar to finding a way to re-align all isomorphic cycles according to a common starting vertex. Fortunately, this is possible according to the notion of window width, previously introduced and expressed by functions W and M. Indeed, the window width discriminates the positions in a sequence, by identifying the ones with the highest local balance, that is to say the ones from which starts the longest minimal sub-sequence containing all values in $\{1, .., n\}$. Obviously, the window width is the same for all isomorphic cycles, as they have the same sequence of local balances up to a cyclic shift, whatever the labels of the dimensions. For any class of cycles, there is at least one position corresponding to the window width and we use it as the reference starting position to force the alignment of all cycles in the class.

When there is exactly one such position in a class, there is no ambiguity and every cycle of the class if shifted to begin at this position. However, for some classes, there might exist several positions corresponding to the window width. Thus, an additional deterministic selection must be applied to those possibilities. This is where the total lexicographic order is exploited, by selecting the position whose ordered relabeling produces the smallest sequence relatively to that order. This is what is expressed by the min operator in function C. As the result is a minimal value over a totally ordered space, it is unique and it ensures the common re-alignment of all the cycles in a same class.

So, function C re-aligns isomorphic cycles to a common starting position and relabels their dimensions in an ordered way that ensures a unique result for isomorphic cycles. □

Finally, it is worth noticing that it is the use of the window width notion combined to cyclic shifts, the total lexicographic order over H_N and the dimensions relabeling that allows us to compute a unique class representative.

So, the binary relation E induced by function C:

$$\forall a, b \in H_N, \quad E(a, b) = \begin{cases} 1 \text{ if } C(a) = C(b) \\ 0 \text{ otherwise} \end{cases} \tag{7}$$

is an equivalence relation over H_N since C is a function.

2.4   Examples of Application of C

Applying function C to our example sequence $h = (2, 3, 1, 3, 2, 3, 1, 3)$ yields $C(h) = (1, 2, 1, 3, 1, 2, 1, 3)$ and $M(h) = 4$. Moreover, that maximal value is obtained for the positions: 2, 4, 6 and 8. So, it is necessary to compute the sequences $R(D(h, 2)), R(D(h, 4)), R(D(h, 6))$, and $R(D(h, 8))$ in order to get the

Table 1. Application of $R \circ D$ to $h = (2, 3, 1, 3, 2, 3, 1, 3)$.

k	$D(h, k)$	$R(D(h, k))$
2	(3,1,3,2,3,1,3,2)	(1,2,1,3,1,2,1,3)
4	(3,2,3,1,3,2,3,1)	(1,2,1,3,1,2,1,3)
6	(3,1,3,2,3,1,3,2)	(1,2,1,3,1,2,1,3)
8	(3,2,3,1,3,2,3,1)	(1,2,1,3,1,2,1,3)

minimal one according to the lexicographic order over H_N, leading to results in Table 1.

Finally, as the four results are identical, the minimal sequence is this unique result, leading to $C(h) = (1, 2, 1, 3, 1, 2, 1, 3)$.

In fact, it can be checked that all the sequences in H_3 lead to that unique result of function C. This comes from the fact that there is only one Gray code up to isomorphism in H_3.

In order to provide a more representative example, let us apply function C to some cycles from a N-cube providing several classes of Hamiltonian cycles. If we consider $h = (1, 2, 1, 3, 4, 3, 2, 1, 2, 3, 4, 2, 4, 1, 4, 3)$ in H_4, then $M(h) = 6$ and there are five possible starting positions: 6, 9, 11, 13 and 16. Figure 3(a) gives the results of $R \circ D$ for those possibilities. The minimal vector is $(1, 2, 1, 3, 1, 4, 3, 2, 3, 4, 1, 4, 2, 3, 2, 4)$, and so is $C(h)$.

Now, let us build a cycle g that is isomorphic to h by applying to h the following operations:

1- invert the sequence order $\quad \rightarrow \quad (3, 4, 1, 4, 2, 4, 3, 2, 1, 2, 3, 4, 3, 1, 2, 1)$
2- apply 4 left cyclic shifts $\quad \rightarrow \quad (2, 4, 3, 2, 1, 2, 3, 4, 3, 1, 2, 1, 3, 4, 1, 4)$
3- exchange dimensions 2 and 3 $\quad \rightarrow \quad (3, 4, 2, 3, 1, 3, 2, 4, 2, 1, 3, 1, 2, 4, 1, 4)$

We thus have $g = (3, 4, 2, 3, 1, 3, 2, 4, 2, 1, 3, 1, 2, 4, 1, 4)$ and $M(g) = 6$ with five corresponding starting positions: 3, 9, 12, 14 and 16. From the application of $R \circ D$ on those instances, depicted in Fig. 3(b), we deduce that $C(g) = (1, 2, 1, 3, 1, 4, 3, 2, 3, 4, 1, 4, 2, 3, 2, 4)$ and that $C(g) = C(h)$.

As a last example, let us consider another cycle in H_4 that is not isomorphic to g and h. This is the case for $f = (3, 1, 4, 1, 2, 1, 4, 1, 3, 1, 4, 1, 2, 1, 4, 1)$ because the numbers of occurrences of the dimensions are not equal in f, whereas they are for g and h. For the computation of $C(f)$, we have $M(f) = 8$ and four corresponding starting positions: 2, 6, 10 and 14. All four positions produce the same result by $R \circ D$, shown in Fig. 3(c), and then $C(f) = (1, 2, 1, 3, 1, 2, 1, 4, 1, 2, 1, 3, 1, 2, 1, 4)$, which is different from $C(g)$ and $C(h)$.

This illustrates the class separation realized by function C when there are several classes in the considered H_N space, as non-isomorphic cycles lead to distinct results whereas isomorphic cycles lead to the same one.

k	$D(h,k)$	$R(D(h,k))$
6	$(3,2,1,2,3,4,2,4,1,4,3,1,2,1,3,4)$	$(1,2,3,2,1,4,1,3,2,3,1,4,3,4,2,4)$
9	$(2,3,4,2,4,1,4,3,1,2,1,3,4,3,2,1)$	$(1,2,3,1,3,4,3,2,4,1,4,2,3,2,1,4)$
11	$(4,2,4,1,4,3,1,2,1,3,4,3,2,1,2,3)$	$(1,2,1,3,1,4,3,2,3,4,1,4,2,3,2,4)$
13	$(4,1,4,3,1,2,1,3,4,3,2,1,2,3,4,2)$	$(1,2,1,3,2,4,2,3,1,3,4,2,4,3,1,4)$
16	$(3,1,2,1,3,4,3,2,1,2,3,4,2,4,1,4)$	$(1,2,3,2,1,4,1,3,2,3,1,4,3,4,2,4)$

(a) $h = (1,2,1,3,4,3,2,1,2,3,4,2,4,1,4,3)$.

k	$D(g,k)$	$R(D(g,k))$
3	$(2,3,1,3,2,4,2,1,3,1,2,4,1,4,3,4)$	$(1,2,3,2,1,4,1,3,2,3,1,4,3,4,2,4)$
9	$(2,1,3,1,2,4,1,4,3,4,2,3,1,3,2,4)$	$(1,2,3,2,1,4,2,4,3,4,1,3,2,3,1,4)$
12	$(1,2,4,1,4,3,4,2,3,1,3,2,4,2,1,3)$	$(1,2,3,1,3,4,3,2,4,1,4,2,3,2,1,4)$
14	$(4,1,4,3,4,2,3,1,3,2,4,2,1,3,1,2)$	$(1,2,1,3,1,4,3,2,3,4,1,4,2,3,2,4)$
16	$(4,3,4,2,3,1,3,2,4,2,1,3,1,2,4,1)$	$(1,2,1,3,2,4,2,3,1,3,4,2,4,3,1,4)$

(b) $g = (3,4,2,3,1,3,2,4,2,1,3,1,2,4,1,4)$.

k	$D(f,k)$	$R(D(f,k))$
2	$(1,4,1,2,1,4,1,3,1,4,1,2,1,4,1,3)$	$(1,2,1,3,1,2,1,4,1,2,1,3,1,2,1,4)$
6	$(1,4,1,3,1,4,1,2,1,4,1,3,1,4,1,2)$	$(1,2,1,3,1,2,1,4,1,2,1,3,1,2,1,4)$
10	$(1,4,1,2,1,4,1,3,1,4,1,2,1,4,1,3)$	$(1,2,1,3,1,2,1,4,1,2,1,3,1,2,1,4)$
14	$(1,4,1,3,1,4,1,2,1,4,1,3,1,4,1,2)$	$(1,2,1,3,1,2,1,4,1,2,1,3,1,2,1,4)$

(c) $f = (3,1,4,1,2,1,4,1,3,1,4,1,2,1,4,1)$

Fig. 3. Application of $R \circ D$ to cycles from H_4.

2.5   Discussion over the Interest of the Canonical Form

This work provides an efficient way to partition the H_N space up to isomorphisms by computing unique representatives of the classes. Such partitions are very useful as soon as one wants to study properties of Gray codes in dimensions larger than 3, as it is possible to focus only on classes representatives. This lead to more efficient algorithms as the number of classes increases much slowly than the number of instances. Moreover, the total order over the class representatives can also be exploited to implement efficient storage and classification algorithms when exploring a given H_N space.

3   Balanced Gray Codes Generation Algorithm

We remind the reader that in balanced Gray codes, the dimensions of the N-cube are used a same number of times or at most with a difference of two occurrences. When all the dimensions are used exactly the same number of times, we speak of totally balanced Gray codes. This is only possible for N-cubes whose dimension is a power of 2.

In order to generate the complete subset of balanced Gray codes in a given H_N space, we have adapted the $(d, g)-$algorithm proposed by Wild [10] to generate all the Hamiltonian cycles. As this algorithm produces more cycles than the ones we are interested in, we had to insert an additional selection phase during the generation process in order to discard branches that would lead to imbalanced Hamiltonian cycles.

That additional selection can be placed before the other treatments (coherency, small cycles elimination,...) applied to each generation node (in the generation tree). By this way, it cuts any unproductive branch as soon as possible, thus avoiding useless computations.

That selection consists in checking that the occurrences of the dimensions already used in the partial construction of the cycle are compatible with a balanced cycle. When this is not the case, the candidate is discarded. To check this, we compute two values that are respectively, the maximal number of occurrences allowed per dimension in a balanced code ($\overline{O}$), and the maximal number of dimensions ($\overline{D}$) with that specific number of occurrences.

Those two numbers can be directly deduced from the dimension N of the N-cube:

$$\overline{O} = \left\lfloor \frac{2^N}{N} \right\rfloor + 2\left(\left\lceil \frac{2^N}{N} \right\rceil - \left\lfloor \frac{2^N}{N} \right\rfloor \right) \text{ and } \overline{D} = \frac{2^N \bmod N}{2} \tag{8}$$

The imbalance detection algorithm is given in Algorithm 1.1.

Algorithm 1.1. Imbalance detection algorithm

```
1 Input: a partially built path p
2 Output: a boolean indicating True if the path is imbalanced and  ↩
                  ↪ False otherwise

4 Initialize array od[] of size n with 0
5 nbD ← 0              // Number of dimensions with max occurrences
6 imb ← False          // We start with balanced path assumption

8 for each valid move in p do
9    get the dimension d along which the move is done
10   od[d] ← od[d] + 1      // move added to occurrences of d
11   if od[d] > O̅ then      // too much moves on dimension d
12      imb ← True          // imbalance
13   else
14      if od[d] = O̅        // dim d reaches max occurrences
15         if nbD = D̅ then  // too much dims with max occs
16            imb ← True    // imbalance
17         else             // new dim with max occs added
18            nbD ← nbD + 1
19         endif
20      endif
21   endif
22 endfor
23 return imb
```

The imbalance is detected as soon as the number of occurrences of one dimension exceeds $\overline{O}$ or the number of dimensions having reached $\overline{O}$ exceeds $\overline{D}$.

Two other algorithmic enhancements may be added to the process. The former is a treatment of the nodes in the generation tree that aims at speeding up the descent towards the leafs, by jumping several levels in the tree in a same iteration. The latter is quite an extension of the former as it consists in starting the generation process not at the root of the tree but several levels deeper. However, experiments show that such additions do not systematically reduce the cost of the algorithm. A deeper study is necessary to precisely determine the impact of those additions.

Finally, all the paths that are totally specified within the generation process (the leaves of the generation tree) are transformed into their canonical form. That form is added to the lexicographically ordered list of balanced Gray codes if not already present.

So, we obtain an algorithm that generates all the non-isomorphic balanced Gray codes in a given H_N space.

4 Application

The first series of experiments is dedicated to the validation of the canonical form previously presented. Then, the second part is dedicated to the balanced Gray code generation algorithm.

4.1 Validation of the Canonical Form and the Generation Algorithm

The first set of experiments consists in checking the completeness of the obtained generation algorithm described in Sect. 3. So, this algorithm is used to experimentally retrieve *all* the classes in N-cubes up to dimension five. For larger dimensions, the number of distinct cycles is too large to be exhaustively computed (777739016577752714 for H_6).

For each set H_N, all Hamiltonian cycles are generated by the algorithm *without* activating the balance selection. Then, canonical forms of the cycles are computed according to C in order to deduce the distinct classes in the space.

The numbers of resulting classes have been compared to the references provided in [1] and initially coming from [5]. Our algorithm has successfully found a unique class for dimensions 2, and 3. It found 9 classes for dimension 4 and 237675 classes for dimension 5. These results confirm the completeness of the generation algorithm.

4.2 Application of the Balanced Gray Code Generation Algorithm

In theory, the presented algorithm can generate all the balanced cycles for a given dimension of N-cube. However, this is not pertinent in practice due to the exponential increase of the number of cycles. In such case, any algorithm would be confronted to two limitations: memory and execution time. For example, our algorithm can generate all the balanced Gray codes for dimensions up to 5 in a few seconds whereas it would take non reasonable time to generate all the cycles for dimensions 6 and above.

Indeed, in our application context of PRNGs, we need only to generate some particular balanced cycles, according to the regarded properties. It is then possible to restrict the search to some particular cycles. So, it should be possible to obtain a fast algorithm for generating specific balanced Gray codes.

Table 2. The 2 balanced cycles generated by e-RB method in dimension 5 and their corresponding mixing time when ε is 10^{-6}.

Num	Sequence of traversed dimensions of the N-cube	Local balance	Mixing time
19708	1 2 3 1 4 1 3 2 3 4 1 5 4 5 3 5 1 2 3 2 4 2 1 4 3 5 4 5 1 2 1 5	12	31
20904	1 2 3 2 1 2 3 4 3 2 1 5 1 4 1 2 4 5 3 5 4 2 4 5 1 5 2 3 4 3 2 5	12	31

Table 3. Excerpt of the 26155 non isomorphic Hamiltonian cycles generated by our method with either the smallest local balance or the smallest mixing time with $\varepsilon = 10^{-6}$ for dimension N = 5.

Num	Sequence of traversed dimensions of the N-cube	Local balance	Mixing time
22534	1 2 3 2 1 4 5 4 1 3 2 3 1 5 4 5 1 2 3 2 1 4 5 4 1 3 2 3 1 5 4 5	7	34
962	1 2 1 3 1 2 4 1 4 5 2 4 5 3 2 1 5 4 3 2 3 1 4 1 5 3 5 2 1 3 4 5	10	29
983	1 2 1 3 1 2 4 1 4 5 4 3 2 1 5 3 5 2 4 2 3 1 3 4 5 3 2 1 5 4 1 5	10	29
8962	1 2 1 3 2 4 2 1 4 5 3 2 4 1 5 1 3 4 5 4 1 2 1 3 5 1 5 4 3 2 3 5	10	29
22624	1 2 3 2 4 1 2 3 4 5 3 5 2 5 1 4 1 2 3 2 5 1 2 3 5 4 3 4 2 4 1 5	10	29
24059	1 2 3 2 4 3 2 1 4 5 3 5 2 5 1 4 1 2 3 2 5 3 2 1 5 4 3 4 2 4 1 5	10	29
11087	1 2 1 3 4 1 2 3 2 1 5 4 3 1 5 2 5 4 1 2 1 3 4 2 4 1 5 3 4 5 3 5	11	29
18407	1 2 3 1 3 4 2 3 2 1 5 3 5 2 5 4 2 1 3 1 2 4 5 4 1 5 4 2 3 2 4 5	11	29
772	1 2 1 3 1 2 1 4 3 2 3 5 2 4 2 3 4 1 5 3 5 4 3 2 3 4 5 3 5 1 4 5	12	29
6759	1 2 1 3 2 1 4 1 3 2 3 5 4 3 4 1 5 2 5 3 2 5 4 5 3 2 3 1 4 3 4 5	12	29
14967	1 2 1 3 4 3 2 1 3 1 4 5 2 1 5 2 5 4 2 5 3 5 4 1 4 3 2 3 1 4 1 5	12	29
16317	1 2 3 1 2 1 4 3 1 3 2 5 1 5 3 5 4 2 4 3 4 2 1 3 5 4 5 2 3 4 3 5	12	29
17396	1 2 3 1 3 4 1 2 1 3 4 5 2 3 4 2 3 2 5 1 5 3 4 3 2 5 4 5 3 4 1 5	12	29

Moreover, compared to other methods to generate balanced Gray codes, like the extended Robinson-Cohn (further denoted as e-RB) algorithm [9] or the Bykov's one [1], our approach presents the advantage of being more complete, and thus more flexible. It is able to find any balanced cycle that has some specific properties, namely which is locally balanced and whose mixing time (time until the Markov chain is ε close to the uniform distribution) is reduced.

As a first example, if we consider dimension 5, the e-RB method can only generate 2 balanced cycles (modulo cycle isomorphism), given in Table 2. The cycles are given in canonical form and the numbers in the left column correspond to their positions in the totally ordered set of all balanced cycles for dimension 5 (26155). Both cycles have a local balance of 12 and a mixing time of 31 where ε is 10^{-6}. However, for this dimension, the minimal local balance is 7 (only one cycle) and the best mixing time is 29 (several cycles with different local balances). All those cycles are listed in Table 3, together with their local balance and mixing time. So, it is clear that our method is better suited to find cycles of interest for the construction of PRNGs.

A second example is related to the Bykov's construction of locally balanced cycles. The proposed algorithm builds a family of Hamiltonian cycles in a N-cube with a specific local balance of at most $n + 3.log_2(n)$. However, Table 3 shows us two facts. The former is that only two cycles among the 7403 with this particular local balance (11 for dimension 5) obtain the minimal mixing time. The latter is that this minimum is reached also by some cycles with other local balances (10 and 12). Thus, a more exhaustive algorithm, like the one we propose, is useful to get all the cycles better suited to the inclusion in a PRNG and to provide a wider choice.

5   Conclusion

A canonical form has been proposed to provide unique representations of Hamiltonian cycles in N-cubes. All the properties of an equivalence relation over the set H_N have been proved. Based on this form and the Wild's algorithm that generates cycles in graphs, a new algorithm has been designed to generate all the balanced Hamiltonian cycles in any N-cube. Restrictions to specific cycles can be used to limit the generation and to avoid the combinatorial explosion on the number of cycles for dimensions greater than 6.

In the application context of PRNG construction, we have shown that our algorithm is better suited than other existing methods that generate only specific cycles, like the extended Robinson-Cohn and the Bykov ones.

Hence, our algorithm provides a useful tool to study the cycles properties that are relevant to the inclusion in a PRNG. This study is planned as our next work, together with performance optimization of our generation algorithm.

Acknowledgments. This work is partially funded by the Labex ACTION program (contract ANR-11-LABX-01-01).

References

1. Bykov, I.S.: On locally balanced gray codes. J. Appl. Ind. Math. **10**(1), 78–85 (2016). http://dx.doi.org/10.1134/S1990478916010099
2. Cao, L., Min, L., Zang, H.: A chaos-based pseudorandom number generator and performance analysis. In: International Conference on Computational Intelligence and Security, CIS 2009, vol. 1, pp. 494–498. IEEE, December 2009
3. Couchot, J.F., Contassot-Vivier, S., Héam, P.C., Guyeux, C.: Random walk in a N-cube without Hamiltonian cycle to chaotic pseudorandom number generation: theoretical and practical considerations. International Journal of Bifurcation and Chaos (2016). Accepted on October 2016
4. Couchot, J., Héam, P., Guyeux, C., Wang, Q., Bahi, J.M.: Pseudorandom number generators with balanced gray codes. In: SECRYPT 2014 - Proceedings of the 11th International Conference on Security and Cryptography, Vienna, Austria, 28–30 August 2014, pp. 469–475 (2014)
5. Sloane, N.: On-line encyclopedia of integer sequences. http://oeis.org

6. Stojanovski, T., Kocarev, L.: Chaos-based random number generators-part i: analysis [cryptography]. IEEE Trans. Circ. Syst. I: Fundam. Theor. Appl. **48**(3), 281–288 (2001)
7. Stojanovski, T., Pihl, J., Kocarev, L.: Chaos-based random number generators part ii: practical realization. IEEE Trans. Circ. Syst. I: Fundam. Theor. Appl. **48**(3), 382–385 (2001)
8. Suparta, I., van Zanten, A.: Totally balanced and exponentially balanced gray codes. Discrete Anal. Oper. Res. (Russia) **11**(4), 81–98 (2004)
9. Suparta, I., van Zanten, A.: A construction of gray codes inducing complete graphs. Discrete Math. **308**(18), 4124–4132 (2008)
10. Wild, M.: Generating all cycles, Chordless cycles, and Hamiltonian cycles with the principle of exclusion. J. Discrete Algorithms **6**(1), 93–103 (2008). http://www.sciencedirect.com/science/article/pii/S1570866707000020, Selected papers from AWOCA 2005 Sixteenth Australasian Workshop on Combinatorial Algorithms

Equicontinuity and Sensitivity
of Nondeterministic Cellular Automata

Pietro Di Lena[(✉)]

Department of Computer Science and Engineering,
University of Bologna, Bologna, Italy
pietro.dilena@unibo.it

Abstract. Nondeterministic Cellular Automata (NCA) are the class
of multivalued functions characterized by nondeterministic block maps.
We extend the notions of equicontinuity and sensitivity to multivalued
functions and investigate the characteristics of equicontinuous, almost
equicontinuous and sensitive NCA. The dynamical behavior of nondeterministic CA in these classes is much less constrained than in the
deterministic setting. In particular, we show that there are transitive
NCA with equicontinuous points and equicontinuous NCA that are not
reversible.

Keywords: Nondeterministic cellular automata · Equicontinuity ·
Sensitivity · Transitivity

1 Introduction

Cellular Automata (CA) are discrete dynamical systems on the space of dou-
bly infinite grid of cells. At any temporal instant, each cell can be in one of a
finite number of possible states. The state of every cell is updated synchronously
according to some fixed *local rule* that depends on the current state of the cell
and that of its neighboring cells. CA represent also one of the simplest abstract
models for parallel computation. The dynamical [3,6,10,17] and computational
[7,9,11,12] properties of the CA formalism, as well as of its asynchronous and
non-uniform variants [4,5,8], have been well studied in literature.

Cellular Automata can be easily extended to nondeterminism by simply
allowing a nondeterministic local rule. Nondeterminism is an important notion
in Computation Theory, hence Nondeterministic Cellular Automata (NCA) represent a natural model for nondeterministic parallel computation. Despite its
attractiveness, so far NCA received very little attention in literature. This may
be due to the fact that NCA are a special class of multivalued functions and
there is a substantial lack of mathematical background for studying multivalued
dynamical systems [1].

The first mention of NCA in literature traces back to the seventies [20,21],
and then there is a gap on the subject until very recently [2,13–15,19]. In [13]

© IFIP International Federation for Information Processing 2017
Published by Springer International Publishing AG 2017. All Rights Reserved
A. Dennunzio et al. (Eds.): AUTOMATA 2017, LNCS 10248, pp. 81–96, 2017.
DOI: 10.1007/978-3-319-58631-1_7

we started to study the most basic properties of the NCA mappings. In particular, we proved necessary and sufficient conditions that characterize the class of nondeterministic block mappings.

In [13] we focused essentially on surjective and reversible NCA, i.e. nondeterministic block mappings whose inverse can be still defined by a nondeterministic block map. In this work we continue our investigation on NCA from the dynamical point of view. We extend some widely studied topological properties, such as equicontinuity and sensitivity, to nondeterministic mappings. These two properties do not have a standardized definition in the multivalued setting. We show several examples of equicontinuous, almost equicontinuous and sensitive NCA. In comparison to CA, the dynamical behavior of NCA is much more complex and less constrained. The largest differences are probably found in the class of equicontinuous NCA. Surjective and equicontinuous CA have a strongly periodic behavior, hence they are bijective and reversible. On the contrary, there are equicontinuous NCA that are transitive and not reversible. Several questions about the NCA dynamics are open. In particular, it is an open question whether there are NCA whose set of equicontinuous points is not empty and non dense, which would imply that there are not sensitive and not almost equicontinuous NCA.

The paper is organized as follows. In Sect. 2 we introduce the basic notation and background. Sections 3 and 4, are devoted to equicontinuity and sensitivity, respectively. In Sect. 5 we consider transitive NCA. Section 6 contains the final remarks.

2   Preliminaries

2.1   Cellular Automata

We introduce the basic notation and terminology we will use throughout the rest of the paper. We assume that the reader is familiar with the elementary notions from Symbolic Dynamics and Topology Theory [16,18].

Let A be a finite set with at least two elements. We denote with A^k, the set of words over A of length $k > 0$, with $A^+ = \cup_{k>0} A^k$ the set of finite words on A and with A^* the set of finite words of A, including the empty word. The set $A^{\mathbb{Z}}$ denotes the set of doubly infinite sequences $(x_i)_{i \in \mathbb{Z}}$ of symbols $x_i \in A$. Given $x \in A^{\mathbb{Z}}$ we use the shortcut $x_{[i,j]}$ for the sub-word $x_i x_{i+1}..x_j \in A^{j-i+1}$. A sequence containing a periodic repetition of the word $w \in A^+$ is denoted with $^{\infty}w^{\infty}$, i.e. $x = {}^{\infty}w^{\infty}$ if $\forall i \in \mathbb{Z}, x_{[i \cdot |w|,(i+1)|w|-1]} = w$.

The mapping $\sigma : A^{\mathbb{Z}} \rightarrow A^{\mathbb{Z}}$, defined by $\sigma(x)_i = x_{i+1}$, is called *shift map*. The pair $(A^{\mathbb{Z}}, \sigma)$ is a dynamical system, called *full shift*.

Consider the metric $d(x,y) = 2^{-n}$ on $A^{\mathbb{Z}}$, where $n = \min\{|i| \mid x_i \neq y_i\}$. The full shift $A^{\mathbb{Z}}$ endowed with metric d is a Cantor space, i.e. a compact, totally disconnected, metric space. For every word $u \in A^+$ and $i \in \mathbb{Z}$, the set $[u]_i = \{x \in A^{\mathbb{Z}} \mid x_{[i,|u|-1]} = u\}$ is called *cylinder set*. A cylinder set is a clopen (closed and open) set in $A^{\mathbb{Z}}$. Given $x \in A^{\mathbb{Z}}$ and $\epsilon = 2^{-r} > 0$, the open ball

$\mathcal{B}_\epsilon(x) = \{y \in A^{\mathbb{Z}} \mid d(x,y) < \epsilon\}$ coincides with the cylinder set $[x_{[-r,r]}]_{-r}$. Every open set $U \subseteq A^{\mathbb{Z}}$ is defined by a countable union of cylinders.

An endomorphism $F : A^{\mathbb{Z}} \to A^{\mathbb{Z}}$ is a *sliding block code* if there exists a *block map* $f : A^{2r+1} \to A$, for some *radius* $r \geq 0$, such that for every point $x \in A^{\mathbb{Z}}$, $F(x)_i = f(x_{[i-r,i+r]})$. We call f *local rule* of F. The fundamental Theorem of symbolic dynamics [16], states that a mapping $F : A^{\mathbb{Z}} \to A^{\mathbb{Z}}$ is a sliding block code if and only if F is *continuous* and *commutes with the shift*, i.e. $F(\sigma(x)) = \sigma(F(x))$. The shift map σ itself is a sliding block code. The continuous and σ-commuting endomorphisms of the full shift $(A^{\mathbb{Z}}, F)$ are usually known as *Cellular Automata* (CA).

2.2 Nondeterministic Cellular Automata

Nondeterministic Cellular Automata (NCA) are the class of *multivalued functions* (or *multimaps*) definable by *nondeterministic block maps*.

Definition 1. *Let A be some alphabet A with at least two elements.*

– *A mutivalued mapping $f : A^{2r+1} \rightrightarrows A$ of radius $r \geq 0$ is a nondeterministic block map if,*

$$\forall w \in A^{2r+1}, \ f(w) \subseteq A$$

– *A multivalued function $F : A^{\mathbb{Z}} \rightrightarrows A^{\mathbb{Z}}$ is a nondeterministic cellular automaton if there is some nondeterministic block map $f : A^{2r+1} \rightrightarrows A$ such that:*

$$\forall x \in A^{\mathbb{Z}}, F(x) = \{y \in A^{\mathbb{Z}} \mid \forall i \in \mathbb{Z}, y_i \in f(x_{[i-r,i+r]})\}$$

Continuity notion for (single-valued) functions can be extended to multivalued functions by means of the dual concepts of *upper* and *lower semicontinuity* (also referred to as *upper* and *lower hemicontinuity*), which collapse to the ordinary notion of continuity in the single-valued setting. The upper and lower semicontinuity properties have a simple characterization in terms of preimages of closed and open sets.

Definition 2. *Let $F : A^{\mathbb{Z}} \rightrightarrows A^{\mathbb{Z}}$ be a multimap.*

– *F is said upper semicontinuous at $x \in A^{\mathbb{Z}}$ if for any open subset $V \subseteq A^{\mathbb{Z}}$ such that $F(x) \subseteq V$,*

$$\exists \delta > 0 \text{ such that } \forall x' \in \mathcal{B}_\delta(x), F(x') \subseteq V$$

– *F is said lower semicontinuous at $x \in A^{\mathbb{Z}}$ if for any open subset $V \subseteq A^{\mathbb{Z}}$ such that $F(x) \cap V \neq \emptyset$,*

$$\exists \delta > 0 \text{ such that } \forall x' \in \mathcal{B}_\delta(x), F(x') \cap V \neq \emptyset$$

– *F is said continuous if it is both lower and upper semicontinuous at every $x \in A^{\mathbb{Z}}$.*   □

Proposition 1. *Let $F : A^{\mathbb{Z}} \rightrightarrows A^{\mathbb{Z}}$ be a multimap.*

1. *F is upper semicontinuous if and only if for any closed set $V \subseteq A^{\mathbb{Z}}$, $F^{-1}(V)$ is closed in $A^{\mathbb{Z}}$.*
2. *F is lower semicontinuous if and only if for any open set $V \subseteq A^{\mathbb{Z}}$, $F^{-1}(V)$ is open in $A^{\mathbb{Z}}$.*

It is easy to prove that nondeterministic block mappings are σ-commuting and continuous. However, these two properties alone are not sufficient to characterize the class of multi-valued functions definable by nondeterministic block maps.

Definition 3. *Let $F : A^{\mathbb{Z}} \rightrightarrows A^{\mathbb{Z}}$ be a multimap.*

- *We say that F is locally independent at $x \in A^{\mathbb{Z}}$ if*

$$y \notin F(x) \text{ if and only if } \exists i \in \mathbb{Z} \text{ such that } y_i \notin F(x)_i$$

  where
$$F(x)_i = \{a \in A \mid \exists z \in F(x), z_i = a\}$$
- *We say that F is locally independent if it is locally independent at every $x \in A^{\mathbb{Z}}$.* □

Theorem 1 [13]. *A multimap $F : A^{\mathbb{Z}} \rightrightarrows A^{\mathbb{Z}}$ is a NCA if and only if it is continuous, σ-commuting and locally independent.*

It is not generally true for multivalued functions that the continuous image of a compact set is compact. It is possible to prove that this property holds for nondeterministic block mappings.

Theorem 2 [13]. *Let $(A^{\mathbb{Z}}, F)$ be a NCA. Then $F(U)$ is compact for every compact subset $U \subseteq A^{\mathbb{Z}}$.*

An interesting class of NCA is the class of *reversible* NCA.

Definition 4. *Let $(A^{\mathbb{Z}}, F)$ be a NCA.*

- *The reversed map $F^{-1} : F(A^{\mathbb{Z}}) \rightrightarrows A^{\mathbb{Z}}$ is defined by*

$$F^{-1}(x) = \{y \in A^{\mathbb{Z}} \mid x \in F(y)\}$$
- *We say that F is a reversible NCA if F^{-1} is defined by a nondeterministic block map.* □

In the deterministic setting reversibility coincides with the injectivity property. In the nondeterministic setting, the scenario is more complex.

Definition 5. *Let $(A^{\mathbb{Z}}, F)$ be a NCA. We say that F is injective if*

$$\forall x, y \in A^{\mathbb{Z}}, x \neq y, F(x) \cap F(y) = \emptyset.$$

□

In [13] we showed that there are no injective NCA other than the class of injective CA. Moreover, if a CA $(A^{\mathbb{Z}}, F)$ is reversible, then $(A^{\mathbb{Z}}, F^{-1})$ is an injective CA, which implies that if $(A^{\mathbb{Z}}, F)$ is strictly nondeterministic, surjective and reversible, then $(A^{\mathbb{Z}}, F^{-1})$ is again strictly nondeterministic. A further characteristic of reversible NCA is that they don't need to be surjective. The simplest non trivial examples of (surjective or not surjective) reversible NCA are the class of NCA with radius zero.

3   Equicontinuity

Equicontinuous dynamical systems are characterized by the presence of *stable points*, called *equicontinuity points*, under the iterations of the map. Dynamical systems whose set of equicontinuity points is dense (residual), are called *almost equicontinuous*. In the CA dynamical systems, the set of equicontinuity points is either dense or empty, and it is inversely invariant. We show an example of NCA whose set of equicontinuous points is not inversely invariant, while it is an open question whether there are NCA that have a non empty and non dense set of equicontinuous points. Furthermore, it is well known that surjective equicontinuous CA are injective, hence bijective. We show an example of surjective, equicontinuous NCA that is not reversible.

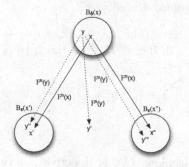 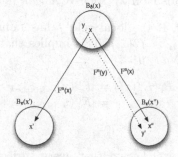

Fig. 1. lower equicontinuous point - upper sensitive

Fig. 2. upper equicontinuous point - lower sensitive

A point x is called equicontinuous if the family of iterations $(F^n)_{n \geq 0}$ is equicontinuous at x. As for the continuity notion for multivalued functions, the dual properties of *lower equicontinuity* and *upper equicontinuity* facilitate the extension of equicontinuity to iterations of the multimap.

Definition 6. *Let $F : A^{\mathbb{Z}} \rightrightarrows A^{\mathbb{Z}}$ be a NCA.*

- *We say that $x \in A^{\mathbb{Z}}$ is an* **upper equicontinuous** *point (Fig. 2) if*

$$\forall \epsilon > 0, \exists \delta > 0 \, such \, that \, \forall y \in B_{\delta}(x), \forall n \geq 0, F^n(y) \subseteq B_{\epsilon}(F^n(x))$$

- *We say that $x \in A^{\mathbb{Z}}$ is a* **lower equicontinuous** *point (Fig. 1) if*

$$\forall \epsilon > 0, \exists \delta > 0 \, such \, that \, \forall y \in B_{\delta}(x), \forall n \geq 0, F^n(x) \subseteq B_{\epsilon}(F^n(y))$$

We say that $x \in A^{\mathbb{Z}}$ is an **equicontinuous** *point if it is both upper and lower equicontinuous:*

$$\forall \epsilon > 0, \exists \delta > 0 \, such \, that \, \forall y \in B_{\delta}(x), \forall n \geq 0, B_{\epsilon}(F^n(x)) = B_{\epsilon}(F^n(y))$$

□

Definition 7. *Let $\mathcal{E} \subseteq A^{\mathbb{Z}}$ be the set of* **equicontinuous points** *of $(A^{\mathbb{Z}}, F)$.*

- *We say that $(A^{\mathbb{Z}}, F)$ is* **equicontinuous** *if $\mathcal{E} = A^{\mathbb{Z}}$.*
- *We say that $(A^{\mathbb{Z}}, F)$ is* **almost equicontinuous** *if $\mathcal{E}$ is a residual set.* □

The most simple class of equicontinuous CA is the class of mappings defined by local rules of radius zero. Such class can be easily characterized also for NCA.

Proposition 2. *Any NCA with radius zero is equicontinuous.*

Proof. If the local rule f has radius zero, we have that $\forall x \in A^{\mathbb{Z}}$, $\forall n \geq 0$, $F^n(x)_i = f^n(x_i)$. This implies that $\forall \epsilon = 2^{-k} > 0$, if $y \in B_{\epsilon}(x) = [x_{[-k,k]}]_{-k}$ then $\forall n \geq 0$

$$B_{\epsilon}(F^n(x)) = [F^n(x)_{[-k,k]}]_{-k} = [f^n(x_{[-k,k]})]_{-k} = [f^n(y_{[-k,k]})]_{-k} =$$
$$= [F^n(y)_{[-k,k]}]_{-k} = B_{\epsilon}(F^n(y))$$

□

It is well known that every surjective equicontinuous CA is injective, hence reversible. We have already shown in [13] that every NCA with radius zero, surjective or not, is reversible. Thus nondeterministic local rules or radius zero give rise to a non trivial class of (either surjective or not) equicontinuous and reversible NCA. However, we can easily show that not every surjective and equicontinuous NCA is reversible.

Example 1 (**Irreversible and equicontinuous NCA**). Consider the NCA $(A^{\mathbb{Z}}, F)$ on the alphabet $A = \{0, 1\}$, defined by the following local rule:

$$\forall a, b, c \in A, f(a, b, c) = \begin{cases} \{0, 1\} \text{ if } a = 1, b = 0, c = 1 \\ \{b\} \quad \text{otherwise} \end{cases}$$

The mapping F is essentially the identity on A, except for the word 101, which is mapped nondeterministically by the local rule to $\{0, 1\}$. F is clearly surjective, since $\forall x \in A^{\mathbb{Z}}, x \in F(x)$. In order to see that $(A^{\mathbb{Z}}, F)$ is equicontinuous, note that 1 is a *quiescent symbol*, i.e. $\forall x \in A^{\mathbb{Z}}, \forall n \geq 0$ if $x_i = 1$ then $F^n(x)_i = \{1\}$. The symbol 0 is quiescent everywhere except when it is immediately surrounded by two (quiescent) 1s. Then,

$$\forall w \in A^3, \forall x, y \in [w]_{-1}, \forall n \geq 0, F^n(x)_0 = F^n(y)_0$$

and, generalizing,

$$\forall w \in A^{2k+1}, \forall x, y \in [w]_{-1}, \forall n \geq 0, F^n(x)_{[-k+1,k-1]} = F^n(y)_{[-k+1,k-1]}$$

which implies equicontinuity. We conclude by showing that F is not reversible. Consider the configuration $\tilde{x} = {}^{\infty}1^{\infty}$, and note that $\tilde{x} \in F({}^{\infty}(01)^{\infty})$ and $\tilde{x} \notin F({}^{\infty}0^{\infty})$. Now, if F^{-1} is defined by some nondeterministic block map $f^{-1} : A^{2k+1} \rightrightarrows A$, the only possibility is that $f^{-1}(1^{2k+1}) = \{0, 1\}$. But in this way, $F^{-1}(\tilde{x}) = A^{\mathbb{Z}}$, while ${}^{\infty}0^{\infty} \notin F^{-1}(\tilde{x})$. $\square$

In topological (single-valued) dynamical systems, the set of equicontinuous points is inversely invariant. This property does not hold for multivalued mappings. The following example shows an almost equicontinuous NCA whose set of equicontinuous points is not invariant.

Example 2 (**Almost equicontinuous NCA with not inversely invariant equicontinuous points**). Consider the NCA $(A^{\mathbb{Z}}, F)$ on alphabet $A = \{0, 1, 2\}$, defined by the following nondeterministic local rule:

$$\forall a, b, c \in A, f(a, b, c) = \begin{cases} \{2\} & \text{if } a = 2 \text{ or } b = 2 \text{ or } c = 2 \\ \{0, 2\} & \text{if } a = b = c = 0 \\ \{c\} & \text{otherwise} \end{cases}$$

Note that, the symbol 2 is a quiescent symbol that spreads to the left and to the right. The point ${}^{\infty}2^{\infty}$ is thus an equicontinuous point of $(A^{\mathbb{Z}}, F)$. Consider the set of sequences that contain infinitely many occurrences of the symbol 2 to the left and to the right.

$$U = \{x \in A^{\mathbb{Z}} \mid \forall i \in \mathbb{N}, \exists k' \geq i, k'' \leq -i, \text{ such that } x_{k'} = 2, x_{k''} = 2\}$$

It is easy to prove (exactly the same proof as in the deterministic case) that U is residual and contains equicontinuous points of $(A^{\mathbb{Z}}, F)$, i.e. $U \subseteq \mathcal{E}$. We show that $\tilde{x} = {}^{\infty}0^{\infty}$ is not an equicontinuous point of $(A^{\mathbb{Z}}, F)$. Let $\delta = 2^{-k}, k \geq 0$ and consider the point $y \in B_{\delta} = [0^{2k+1}]_{-k}$ such that

$$y_i = \begin{cases} x_i \text{ if } i \neq k + 1 \\ 1 \text{ if } i = k + 1 \end{cases}$$

Then $F^{k+1}(\tilde{x})_0 = \{0, 2\} \neq \{1, 2\} = F^{k+1}(y)$, which implies that $\tilde{x}$ is not an equicontinuous point and that $(A^{\mathbb{Z}}, F)$ is almost equicontinuous but not equicontinuous. To conclude, note that $\tilde{x} = {}^{\infty}0^{\infty} \in F^{-1}(U)$, but since $\tilde{x} \notin U$ we conclude that $F^{-1}(\mathcal{E}) \not\subseteq \mathcal{E}$. $\square$

In CA equicontinuity is strictly related to the presence of blocking words.

Definition 8. *Let $\epsilon = 2^{-k}, k \geq 0$. A word $w \in A^{2d+1}$, is a blocking word if*

$$\forall x, y \in [w]_{-d}, \forall n \geq 0, F^n(x)_{[-k,k]} = F^n(y)_{[-k,k]}.$$

$\square$

In particular, equicontinuous points of CA are characterized by the presence of infinitely many occurrences of *blocking words*. This strong characterization implies that the set of equicontinuous points of a cellular automaton is either empty or dense (residual). There is no immediate generalization of such property for blocking words of NCA, as shown by the following example. This leaves open the question whether there are NCA whose set of equicontinuous points is non empty and not dense.

Example 3 (**NCA with dense set of equicontinuous points**). Consider the NCA $(A^{\mathbb{Z}}, F)$ on the alphabet $A = \{0, 1\}$, defined by the following nondeterministic local rule

$$\forall a, b, c \in A, f(a, b, c) = \begin{cases} \{0, 1, 2\} & \text{if } b = 2 \\ \{b\} & \text{if } c = 2 \\ \{c\} & \text{otherwise} \end{cases}$$

Note that the function F behaves like the shift map on $\{0, 1\}^{\mathbb{Z}}$ and that for every $x \in A^{\mathbb{Z}}$, $F(x) \cap \{0, 1\}^{\mathbb{Z}} \neq \emptyset$. On the other end, the symbol 2 does not move and generates all the other symbols. We first show that $(A^{\mathbb{Z}}, F)$ has a dense set of equicontinuous points. It is easy to see that $\tilde{x} = {}^{\infty}2^{\infty}$ is an equicontinuous point, since $\forall \epsilon = 2^{-k}, k \geq 0$ and for every $\delta \leq \epsilon$

$$\forall y \in B_{\delta}(\tilde{x}), \forall n > 0, F^n(y)_{[-k,k]} = F^n(\tilde{x})_{[-k,k]} = A^{2k+1}$$

In the same way, for every $w \in A^*$, all the points in

$$U_w = \{x \in A^{\mathbb{Z}} \mid \exists i \in \mathbb{Z}, x_{[i, i+|w|-1]} = w \wedge \forall j \notin [i, i+|w|-1], x_j = 2\}$$

are equicontinuous. Then the dense set $U = \cup_{w \in A^*} U_w$ is contained in $\mathcal{E}$.

Now, fix some $\epsilon = 2^{-k}, k \geq 0$. For simplicity we consider $k = 0$, but what follows can be generalized to larger k. By definition, the word $w = 2^{2k+1} = 2$ is blocking. We show that there are sequences containing infinite occurrences of w to left and to the right that are not equicontinuous. Consider, for example, the periodic sequence $\tilde{x} = {}^{\infty}(0002)^{\infty}$, such that $\tilde{x}_{[-1,1]} = 000$. Since the word 2 cannot generate all the 3-words on $\{0, 1\}$, for $\epsilon = 2^{-1}$ and for every $\delta = 2^{-d}$ we can build the configuration $y \in B_{\delta}(\tilde{x})$ such that $y_{[d+1,d+3]} = 111$ and $y_i = x_i$ for every $i \notin [d+1, d+3]$. Then, it is easy to see that:

$$\exists n > 0, 111 \in F^n(y)_{[-1,1]}, \text{ while } \forall n \geq 0, 111 \notin F^n(\tilde{x})_{[-1,1]}$$

which implies that $\tilde{x}$ is not a point of equicontinuity for F.

$\square$

4   Sensitivity

In sensitive dynamical systems small perturbations of an orbit may lead to significantly different trajectories. In some sense, sensitivity is the opposite of equicontinuity and, in fact, the two notions are strictly related: a sensitive dynamical system cannot have points of equicontinuity. The converse is not generally true, although it is for the CA dynamical systems. The question is open for NCA and it is strictly related to the question whether there is a NCA whose set of equicontinuous points is not empty and not dense.

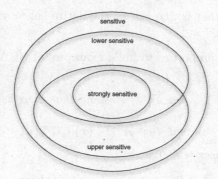

Fig. 3. Sensitivity classes

There is no standard definition of sensitivity for multimaps. We extend the usual definition of sensitivity to multimaps by introducing the notion of upper and lower sensitivity. We get different classes of sensitivity that coincide with the classical definition when the mapping is single-valued.

Definition 9. *Let $F : A^{\mathbb{Z}} \rightrightarrows A^{\mathbb{Z}}$ be a NCA.*

- *We say that $(A^{\mathbb{Z}}, F)$ is* **upper sensitive** *(Fig. 1) if*

$$\exists \epsilon > 0, \forall x \in A^{\mathbb{Z}}, \forall \delta > 0, \exists y \in B_\delta(x), \exists n \geq 0, F^n(y) \not\subset B_\epsilon(F^n(x))$$

- *We say that $(A^{\mathbb{Z}}, F)$ is* **lower sensitive** *(Fig. 2) if*

$$\exists \epsilon > 0, \forall x \in A^{\mathbb{Z}}, \forall \delta > 0, \exists y \in B_\delta(x), \exists n \geq 0, F^n(x) \not\subset B_\epsilon(F^n(y))$$

- *We say that $(A^{\mathbb{Z}}, F)$ is* **sensitive** *if*

$$\exists \epsilon > 0, \forall x \in A^{\mathbb{Z}}, \forall \delta > 0, \exists y \in B_\delta(x), \exists n \geq 0, F^n(y) \not\subset B_\epsilon(F^n(x)) \vee F^n(x) \not\subset B_\epsilon(F^n(y))$$

- *We say that $(A^{\mathbb{Z}}, F)$ is* **strongly sensitive** *if*

$$\exists \epsilon > 0, \forall x \in A^{\mathbb{Z}}, \forall \delta > 0, \exists y \in B_\delta(x), \exists n \geq 0, F^n(y) \cap B_\epsilon(F^n(x)) = \emptyset$$

$\square$

The ϵ constant is called *sensitivity constant* of the map. All four sensitivity classes imply no equicontinuous points. Note that upper and lower sensitivity imply sensitivity but the converse is not immediately false. Strong sensitivity immediately implies lower and upper sensitivity. While in the deterministic setting all four definitions are equivalent, in the nondeterministic setting, the four definitions give rise to different classes of sensitivity. We show that all such classes are non empty and distinct (see Fig. 3).

In the following two lemmas we prove the following properties:

1. If some nondeterministic orbit appears in every other orbit, then the NCA is not lower sensitive,
2. If there is a point that is mapped to the entire space, then the NCA is not upper sensitive.

We will use these two properties to build examples of NCA that are not lower- or-upper sensitive.

Lemma 1. *Let $(A^{\mathbb{Z}}, F)$ be an NCA. Assume that there is some point $x \in A^{\mathbb{Z}}$ such that*

$$\forall n > 0, \forall y \in A^{\mathbb{Z}}, F^n(x) \subseteq F^n(y)$$

then $(A^{\mathbb{Z}}, F)$ is not lower sensitive.

Proof. Let $(A^{\mathbb{Z}}, F)$ be of radius $r \geq 0$ and assume there is one point $x \in A^{\mathbb{Z}}$ as defined in the statement. Consider some $\epsilon > 0$, then

$$\forall \delta > 0, \forall y \in B_{\delta}(x), \forall n > 0, F^n(x) \subseteq F^n(y) \subseteq B_{\epsilon}(F^n(y)),$$

which implies that F is not lower sensitive. $\square$

Lemma 2. *Let $(A^{\mathbb{Z}}, F)$ be an NCA. Assume that there is some point $x \in A^{\mathbb{Z}}$ such that*

$$F(x) = A^{\mathbb{Z}}.$$

Then F in not upper sensitive.

Proof. First of all, note that if $F(x) = A^{\mathbb{Z}}$, then $\forall n > 0, F^n(x) = A^{\mathbb{Z}}$. Consider some $\epsilon > 0$, then

$$\forall \delta > 0, \forall y \in B_{\delta}(x), \forall n > 0, F^n(y) \subseteq B_{\epsilon}(F^n(x)) = B_{\epsilon}(A^{\mathbb{Z}}),$$

which implies that F is not upper sensitive. $\square$

All the following examples are based on the shift map. We first show that sensitivity does not imply lower and upper sensitivity.

Example 4 (**Sensitive but not lower/upper sensitive NCA**). Consider the NCA $(A^{\mathbb{Z}}, F)$ on alphabet $A = \{0, 1\}$ defined by the following nondeterministic local rule:

$$\forall a, b, c \in A, f(a, b, c) = \begin{cases} \{0\} & \text{if } c = 0 \\ \{0, 1\} & \text{if } c = 1 \end{cases}$$

This nondeterministic map contains both the shift map and the constant map, which sends every configuration to the uniform configuration $^\infty 0^\infty$.

We first show that $(A^{\mathbb{Z}}, F)$ is sensitive. Consider some configuration $x \in A^{\mathbb{Z}}$ and note that, by definition of the local rule f, for every $i > 0$

$$F^i(x)_0 = \begin{cases} \{0\} & \text{if } x_i = 0 \\ \{0, 1\} & \text{if } x_i = 1 \end{cases}$$

Let $x \in A^{\mathbb{Z}}$ and let $k > 0$. Let $y \in [x_{[-k,k]}]_{-k}$ be such that

$$y_i = \begin{cases} x_i & \text{if } i \neq k + 1 \\ 1 - x_i & \text{if } i = k + 1 \end{cases}$$

Then $F^{k+1}(x)_0 \neq F^{k+1}(y)_0$, which implies that F is sensitive with sensitivity constant $\epsilon = 2^0$. We now show that F is neither lower nor upper sensitive.

1. F is not upper sensitive. Consider the configuration $\tilde{x} = {}^\infty 1^\infty \in A^{\mathbb{Z}}$. We have that, $\tilde{x}$ is mapped to the entire configuration space, i.e.

$$\forall n > 0, F^n(\tilde{x}) = A^{\mathbb{Z}}.$$

   then, by Lemma 2, F is not upper sensitive.
2. F is not lower sensitive. Consider the configuration $\tilde{x} = {}^\infty 0^\infty \in A^{\mathbb{Z}}$. We have that $\tilde{x}$ is a quiescent configuration that appears in every orbit, i.e.

$$\forall n > 0, \forall y \in A^{\mathbb{Z}}, F^n(\tilde{x}) = \{\tilde{x}\} \subseteq F^n(y).$$

   then, by Lemma 1, F is not lower sensitive.

$\square$

The following two examples show that upper sensitivity does not imply lower sensitivity, and conversely.

Example 5 (**Upper sensitive and not lower sensitive NCA**). Consider the NCA $(A^{\mathbb{Z}}, F)$ on alphabet $A = \{0, 1, 2\}$ defined by the following nondeterministic local rule:

$$\forall a, b, c \in A, f(a, b, c) = \begin{cases} \{0\} & \text{if } c = 0 \\ \{0, 1\} & \text{if } c = 1 \\ \{0, 2\} & \text{if } c = 2 \end{cases}$$

Consider some configuration $x \in A^{\mathbb{Z}}$ and note that, by definition of the local rule f, for every $i > 0$

- if $x_i = 0$, then $F^i(x)_0 = \{0\}$,
- if $x_i = 1$, then $F^i(x)_0 = \{0, 1\}$,
- if $x_i = 2$, then $F^i(x)_0 = \{0, 2\}$,

Then, for every $x \in A^{\mathbb{Z}}$ and $\delta = 2^{-k}, k \geq 0$ we can build the configuration $y \in B_\delta(x) = [x_{[-k,k]}]_{-k}$ such that

$$y_i = \begin{cases} x_i & \text{if } i \neq k+1 \\ 2 & \text{if } i = k+1 \text{ and } x_i \in \{0,1\} \\ 1 & \text{if } i = k+1 \text{ and } x_i = 2 \end{cases}$$

It is clear that $F^{k+1}(y)_0 \not\subset F^{k+1}(x)_0$, which proves that F is upper sensitive with sensitivity constant $\epsilon = 2^0$. In order to see that F is not lower sensitive, consider the uniform configuration $\tilde{x} = {}^\infty 0^\infty \subset A^{\mathbb{Z}}$, which is mapped to itself, i.e. $\forall n \geq 0, F^n(\tilde{x}) = \{\tilde{x}\}$. Note that $\forall y \in A^{\mathbb{Z}}$ and $\forall n > 0, F^n(\tilde{x}) \subseteq F^n(y)$, then by Lemma 1, F is not lower sensitive. $\square$

Example 6 (**Lower sensitive and not upper sensitive NCA**). Consider the one-sided NCA on alphabet $A = \{0,1,2\}$ defined by the following nondeterministic local rule:

$$\forall a, b, c \in A, f(a,b,c) = \begin{cases} \{0,1,2\} & \text{if } a = b = c = 0 \\ \{c\} & \text{otherwise} \end{cases}$$

Note that, for every $x \in A^{\mathbb{Z}}$ and $i > 0$

$$F^i(x)_0 = \begin{cases} \{0,1,2\} & \text{if } x_{[i-2,i]} = 000 \\ \{x_i\} & \text{otherwise} \end{cases}$$

For every $x \in A^{\mathbb{Z}}$ and for every $\delta = 2^{-k}, k \geq 0$ we can build the configuration $y \in B_\delta(x) = [x_{[-k,k]}]_{-k}$ such that

$$y_i = \begin{cases} x_i & \text{if } i \neq k+1 \\ 2 & \text{if } i = k+1 \text{ and } x_i \in \{0,1\} \\ 1 & \text{if } i = k+1 \text{ and } x_i = 2 \end{cases}$$

By construction, we have that, if $F^{k+1}(x) = \{0,1,2\}$ or $F^{k+1}(x) = \{1\}$, then $F^{k+1}(y) = \{2\}$, while if $F^{k+1}(x) = \{1\}$ then $F^{k+1}(y) = \{2\}$. In both cases, $F^{k+1}(x) \not\subset F^{k+1}(y)$, which implies that F is lower sensitive with sensitivity constant $\epsilon = 2^0$. In order to see that F is not upper sensitive, note that the configuration $\tilde{x} = {}^\infty 0^\infty \in A^{\mathbb{Z}}$ is mapped to the entire space, i.e. $F^n(\tilde{x}) = A^{\mathbb{Z}}$, $\forall n > 0$. Then, by Lemma 2, F is not upper sensitive. $\square$

Since CA are a subset of NCA, all sensitive CA belong to the strongly sensitive class. We show that such class contains also strictly NCA. The simples example is the nondeterministic reformulation of the shift map.

Example 7 (**Strongly sensitive NCA**). Consider the NCA $(A^{\mathbb{Z}}, F)$ on alphabet $A = \{0, 1, 2.3\}$ defined by the following nondeterministic local rule:

$$\forall a, b, c \in A, f(a,b,c) = \begin{cases} \{0,2\} & \text{if } c \in \{0,2\} \\ \{1,3\} & \text{if } c \in \{1,3\} \end{cases}$$

Note that, F is essentially a nondeterministic shift map on the two sets $\{0, 2\}$ and $\{1, 3\}$:

$$\forall i > 0, F^i(x)_0 = \begin{cases} \{0, 2\} \text{ if } x_i \in \{0, 2\} \\ \{1, 3\} \text{ if } x_i \in \{1, 3\} \end{cases}$$

For every $x \in A^{\mathbb{Z}}$ and $\delta = 2^{-k}, k \geq 0$ there is the configuration $y \in B_\delta(x)$ such that

$$y_i = \begin{cases} x_i & \text{if } i \neq k + 1 \\ (x_i + 1) \mod 4 & \text{if } i = k + 1 \end{cases}$$

It is easy to see that $F^{k+1}(x)_0 \cap F^{k+1}(y)_0 = \emptyset$, which implies that F is strongly sensitive with sensitivity constant $\epsilon = 2^0$.   □

It is open the question whether there are NCA, both upper and lower sensitive, that are not strongly sensitive.

5 Transitivity

In a *topologically transitive* dynamical system every open set has points whose orbits intersect any other open set. While, in general, a transitive dynamical system can be either sensitive or almost equicontinuous, it is well known that topologically transitive CA are sensitive. This strong characteristic does not hold for NCA.

Definition 10. *A NCA $(A^{\mathbb{Z}}, F)$ is* **topologically transitive** *if for every non-empty open sets $U, V \in A^{\mathbb{Z}}, \exists n \geq 0, F^n(U) \cap V \neq 0$.*

The following general property holds for any continuous endomorphism of a compact space.

Proposition 3. *Any transitive NCA is surjective.*

Proof. Since F is topologically transitive, for every non-empty open set $U \in A^{\mathbb{Z}}$, $F(A^{\mathbb{Z}}) \cap U \neq \emptyset$, which implies that $F(A^{\mathbb{Z}})$ is dense in $A^{\mathbb{Z}}$. Since F is continuous and $A^{\mathbb{Z}}$ compact, $F(A^{\mathbb{Z}})$ is closed, then $F(A^{\mathbb{Z}}) = A^{\mathbb{Z}}$.   □

The following sufficient condition is useful to build examples of transitive NCA.

Definition 11. *Let $(A^{\mathbb{Z}}, F)$ be a NCA with local rule $f : A^{2r+1} \rightrightarrows A$. We say that F' is a sub-NCA of F if its local rule $f' : A^{2r+1} \rightrightarrows A$ is such that*

$$\forall w \in A^{2r+1}, f'(w) \subseteq f(w).$$

If a sub-NCA F' of F is deterministic, we denote it as sub-CA.

Lemma 3. *If there is a sub-NCA F' such that $(A^{\mathbb{Z}}, F')$ is transitive, then $(A^{\mathbb{Z}}, F)$ is transitive.*

Proof. If $(A^{\mathbb{Z}}, F')$ is transitive then, for every non-empty open sets $U, V \subseteq A^{\mathbb{Z}}$ there is $n \geq 0$ such that $F^n(U) \cap V \supseteq (F')^n(U) \cap V \neq \emptyset$.   □

By Lemma 3, all the examples of sensitive NCA in Sect. 4 are transitive, since all of them contain the shift map as sub-CA. However, while it is well known that transitive CA are sensitive, this is not true in the nondeterministic setting. We conclude this section by showing examples of reversible and not reversible transitive NCA that are equicontinuous.

The most simple example of transitive NCA is the map that sends every point into the entire configuration space. Such map is equicontinuous.

Example 8 (**Transitive, reversible and equicontinuous NCA**). Consider the NCA on alphabet $A = \{0, 1\}$ defined by the following nondeterministic local rule of radius zero

$$\forall a \in A, f(a) = A$$

By Proposition 2, $(A^{\mathbb{Z}}, F)$ is equicontinuous. It is clearly transitive, since $\forall n > 0, \forall x \in A^{\mathbb{Z}}$, and for every open set $U \subseteq A^{\mathbb{Z}}, F^n(x) \cap U = A^{\mathbb{Z}} \cap U = U$. This example is also easily reversible and the inverse is the map itself.   □

With a small modification of the previous example, we can get a non-reversible, equicontinuous and transitive NCA.

Example 9 (**Transitive, irreversible and equicontinuous NCA**). Consider the NCA on alphabet $A = \{0, 1\}$ defined by the following nondeterministic local rule:

$$\forall a, b, c \in A, f(a, b, c) = \begin{cases} \{0\} & \text{if } a = c = 1, b = 0 \\ \{0, 1\} & \text{otherwise} \end{cases}$$

By Lemma 3, $(A^{\mathbb{Z}}, F)$ is transitive, since it contains the sensitive elementary rule 90. It is easy to see that it is equicontinuous, since $\forall x \in A^{\mathbb{Z}}, \forall n \geq 2, F^n(x) = A^{\mathbb{Z}}$. In order to see that it is not reversible, consider the configurations $\tilde{x} = {}^{\infty}1^{\infty}$ and $\tilde{y} = {}^{\infty}0^{\infty}$. Note that $\tilde{x} \in F(\tilde{x})$ and $\tilde{x} \in F(\tilde{y})$, thus, if F^{-1} is a nondeterministic block map, the only possibility is that for some $r \geq 0, f^{-1}(1^{2r+1}) = \{0, 1\}$, which implies that $F^{-1}(\tilde{x}) = A^{\mathbb{Z}}$. This is not possible, since $\tilde{x} \notin F({}^{\infty}(01)^{\infty})$.   □

6   Conclusions

We investigated topological dynamical properties of Nondeterministic Cellular Automata. First, we extended to multivalued functions the notions of equicontinuity and sensitivity, which do not have a standard definition as in the single-valued setting. Then, we studied the classes of equicontinuous, almost equicontinuous and sensitive NCA and their intersections with the class of transitive NCA. The topological dynamics of NCA is extremely complex and there are strong differences with respect to their deterministic counterpart. The largest differences are probably found in the class of equicontinuous NCA. Surjective and equicontinuous CA have a strongly periodic behavior, hence they are bijective and reversible. On the contrary, there are equicontinuous NCA that are transitive and not reversible.

There are several interesting open questions. It is unknown whether there is a NCA whose set of equicontinuous point is not empty and not dense. This question is strictly related to the question whether there are NCA that are not sensitive and not almost equicontinuous, which is also open.

References

1. Berge, C.: Topological Spaces. Including a Treatment of Multi-Valued Functions, Vector Spaces and Convexity. Dover Publications Inc., Mineola (1963)
2. Burkhead, E., Hawkins, J.M.: Nondeterministic and stochastic cellular automata and virus dynamics. Preprint
3. Dennunzio, A., Di Lena, P., Formenti, E., Margara, L.: On the directional dynamics of additive cellular automata. Theoret. Comput. Sci. **410**, 4823–4833 (2009)
4. Dennunzio, A., Formenti, E., Manzoni, L., Mauri, G.: m-Asynchronous cellular automata: from fairness to quasi-fairness. Natural Comput. **12**, 561–572 (2013)
5. Dennunzio, A., Formenti, E., Provillard, J.: Local rule distributions, language complexity and non-uniform cellular automata. Theor. Comput. Sci. **504**, 38–51 (2013)
6. Dennunzio, A., Di Lena, P., Formenti, E., Margara, L.: Periodic orbits and dynamical complexity in cellular automata. Fundam. Inform. **126**, 183–199 (2013)
7. Dennunzio, A., Formenti, E., Weiss, M.: Multidimensional cellular automata: closing property, quasi-expansivity, and (un)decidability issues. Theor. Comput. Sci. **516**, 40–59 (2014)
8. Dennunzio, A., Formenti, E., Manzoni, L., Mauri, G., Porreca, A.E.: Computational complexity of finite asynchronous cellular automata. Theor. Comput. Sci. **664**, 131–143 (2017)
9. Di Lena, P.: Decidable properties for regular cellular automata. In: Fourth IFIP International Conference on Theoretical computer Science, pp. 185–196 (2006)
10. Di Lena, P., Margara, L.: Computational complexity of dynamical systems: the case of cellular automata. Inform. Comput. **206**, 1104–1116 (2008)
11. Di Lena, P., Margara, L.: On the undecidability of the limit behavior of Cellular Automata. Theoret. Comput. Sci. **411**, 1075–1084 (2010)
12. Di Lena, P., Margara, L.: On the undecidability of attractor properties for cellular automata. Fund. Inform. **115**, 78–85 (2012)
13. Di Lena, P., Margara, L.: Nondeterministic cellular automata. Inform. Sci. **287**, 13–25 (2014)
14. Furusawa, H., Ishida, T., Kawahara, Y.: Continuous relations and Richardson's Theorem. In: Kahl, W., Griffin, T.G. (eds.) RAMICS 2012. LNCS, vol. 7560, pp. 310–325. Springer, Heidelberg (2012). doi:10.1007/978-3-642-33314-9_21
15. Furusawa, H.: Uniform continuity of relations and nondeterministic cellular automata. Theoret. Comput. Sci. **673**, 19–29 (2017)
16. Hedlund, G.A.: Endomorphisms and automorphisms of the shift dynamical system. Math. Syst. Theory **3**, 320–375 (1969)
17. Kůrka, P.: Languages, equicontinuity and attractors in cellular automata. Ergod. Theor. Dyn. Syst. **17**, 417–433 (1997)
18. Kůrka, P.: Topological and Symbolic Dynamics. Cours Spécialisés 11, Société Mathématique de France, Paris (2003)

19. Ozhigov, Y.: Computations on nondeterministic cellular automata. Inform. Comput. **148**, 181–201 (1999)
20. Richardson, D.: Tessellations with local transformations. J. Comput. Syst. Sci. **6**, 373–388 (1972)
21. Yaku, T.: Surjectivity of nondeterministic parallel maps induced by nondeterministic cellular automata. J. Comput. Syst. Sci. **12**, 1–5 (1976)

Diploid Cellular Automata:
First Experiments on the Random Mixtures of Two Elementary Rules

Nazim Fatès[✉]

Inria Nancy – Grand Est; LORIA UMR 7503, Villers-lès-Nancy, France
nazim.fates@loria.fr

Abstract. We study a small part of the 8088 diploid cellular automata. These rules are obtained with a random mixture of two deterministic Elementary Cellular Automata. We use numerical simulations to study the mixtures obtained with three blind rules: the null rule, the identity rule and the inversion rule. As the mathematical analysis of such systems is a difficult task, we use numerical simulations to get insights into the dynamics of this class of stochastic cellular automata. We are particularly interested in studying phase transitions and various types of symmetry breaking.

Keywords: Stochastic cellular automata · Probabilistic cellular automata · Symmetry breaking · Synchronisation

1 Introduction

As introduced by Turing in his article on morphogenesis [15], the question of randomness is fundamental to understand the laws of life. Turing was puzzled by the capacity of biological organisms to realise a form of "symmetry breaking": given a dynamical system which is invariant under some symmetry and an initial condition which is also invariant under this symmetry, such systems evolve to a state which is stable and which is no longer invariant under this symmetry. The typical case is the embryo which initially has a spherical form and needs to "break" this form to develop different organs. To produce such a phenomenon, two mechanisms can be at play: (a) the initial condition possesses a small asymmetry which is amplified in the evolution of the system; (b) the components of the system evolve with some randomness, which allows them to "choose" one of the symmetric directions of evolutions.

The framework of stochastic cellular automata is adapted for the study of such phenomena. In this note, we deal with one-dimensional cellular automata with random transitions, binary states and nearest neighbours. More specifically, we restrict our scope to the rules defined as a random mixture of two *deterministic* Elementary cellular automata (see the definitions below). As we will see,

A. Dennunzio et al. (Eds.): AUTOMATA 2017, LNCS 10248, pp. 97–108, 2017.
DOI: 10.1007/978-3-319-58631-1_8

this case is sufficiently rich to provide many worthy examples, with potential applications in the study of physical, chemical or biological systems. We are particularly interested in the *qualitative transformations* that a cellular system may undergo when one progressively varies the mixing rate of its two deterministic components.

2   Definitions

Elementary Cellular Automata. We consider finite cellular automata with binary states and periodic boundary conditions. We denote by $Q = \{0, 1\}$ the set of states and let $\mathcal{L} = \mathbb{Z}/n\mathbb{Z}$ denote the set of n cells arranged in a ring. A configuration represents the collection of states of all the cells at a given time step; the set of configurations is denoted by $Q^{\mathcal{L}}$.

In this note, we are only interested in the systems where the cells update their state according to their own state and the state of their left and right neighbour. This set of three cells forms the *neighbourhood* of a given cell. At each time step, the updates are made synchronously according to a *local rule* $f : Q^3 \to Q$. Given a local function f and a set of cells $\mathcal{L}$, one can define the Elementary Cellular Automaton $F : Q^{\mathcal{L}} \to Q^{\mathcal{L}}$ such that, the image $y = F(x)$ of a configuration $x \in Q^{\mathcal{L}}$ is given by:

$$\forall i \in \mathcal{L},\ y_i = f(x_{i-1}, x_i, x_{i+1}).$$

An initial condition $x \in Q^{\mathcal{L}}$ is then associated to its *trajectory*, that is, the infinite sequence $(x^t)_{t \in \mathbb{N}}$ such that $x^0 = x$ and $x^{t+1} = F(x)$ for all $t \in \mathbb{N}$.

Classically, a rule f is identified by the decimal number W which results from the conversion of the binary number $[f(1, 1, 1)f(1, 1, 0) \ldots f(0, 0, 0)]_b$. (In other words $W(f) = f(0, 0, 0) \cdot 2^0 + f(0, 0, 1) \cdot 2^1 + \cdots + f(1, 1, 1) \cdot 2^7$.) It is common to identify the rule with its decimal code. When the ring size n and the local rule f are clear from the context, it is also common to identify the local rule with the Elementary Cellular Automaton (ECA) F.

We now turn our attention to stochastic cellular automata[1]. In this model, each cell has a probability to turn to 0 or to 1 according to the states of its neighbourhood. This means that we need to replace the trajectories $(x^t)_{t \in \mathbb{N}}$ by a *random process* $(\xi^t)_{t \in \mathbb{N}}$. Each ξ^t is formed by the random variables $(\xi_i^t)_{i \in \mathcal{L}}$, which represent the random state of the cell i at time t. Starting from given probability distribution ξ^0, the evolution of the stochastic cellular automaton is a *Markov chain*: intuitively this means that the knowledge of the state of the system at time t is sufficient to calculate the probability distribution at time $t+1$. Since the state of each cell is calculated independently, we can use a function $\phi : Q^3 \to [0, 1]$, which gives the probability to be in state 1 if the neighbourhood of a cell is (x, y, z). Starting from an initial random distribution ξ^0, the evolution of our stochastic cellular automaton can thus be obtained for each $t \in \mathbb{N}$ and for each $i \in \mathcal{L}$ with:

$$\Pr\{\xi_i^{t+1} = 1\} = \sum_{(x,y,z) \in Q^3} \Pr\{(\xi_{i-1}^t, \xi_i^t, \xi_{i+1}^t) = (x, y, z)\} . \phi(x, y, z) \qquad (1)$$

---

[1] The name *probabilistic cellular automata* is also frequently used and is a synonym.

An alternative way of defining the Markov chain $(x^t)_{t \in \mathbb{N}}$ is to specify, for two configurations $x, y \in Q^{\mathcal{L}}$, the probability to go from x to y; we then have:

$$\Pr\{\xi^{t+1} = y | \xi^t = x\} = \prod_{i \in \mathcal{L}} y_i \cdot \phi(x_{i-1}, x_i, x_{i+1}) + (1 - y_i) \cdot \big(1 - \phi(x_{i-1}, x_i, x_{i+1})\big).$$

The formalism defined above is convenient for binary finite systems but it requires more elaborate definitions for infinite-size systems [8] or systems with more states or continuous states [2]. Note that other types of presentations of the rules may be used to facilitate the analysis [2,5].

Diploid ECA. Since a stochastic ECA ϕ is defined with eight probabilities, it is rather intuitive to try to understand how the combination of these eight probabilities influences the behaviour of a rule, for example if the rule may converge to a fixed point or not, and what is the average time of convergence. An important question is to determine what are the various "behaviours" that this space contains and how these behaviours are modified when one moves *continuously* in this space. Since this space already contains the 256 deterministic ECAs, which are difficult to understand theoretically (see e.g. [12]), one can easily guess that its generalisation to the space of probabilistic CA will generate a great number of new problems [1,8].

In order to get some insights on this question, it is thus necessary to concentrate on a subset of the stochastic ECA space. We thus propose to focus on the stochastic mixtures of two deterministic ECA. Such rules have already been studied by many authors (e.g. [9]), but to our knowledge only in particular contexts. We want here to enlarge our view of this space of rules: as a first step, we examine some of their simple properties and explore their dynamics with numerical simulations.

We denote by $\mathcal{S}_8$ the set of stochastic ECA; this set is isomorphic to $[0,1]^8$ and for the sake of simplicity we will identify a stochastic rule and the function which maps each neighbourhood $(x, y, z) \in Q^3$ to its probability to update to state 1. Let $\phi \in \mathcal{S}_8$ be a stochastic ECA, we say that ϕ is a *randomly-mixed ECA* if there exists two ECAs $f_1 : Q^3 \to Q$ and $f_2 : Q^3 \to Q$ and a constant $\lambda \in [0,1]$, called the *mixing rate*, such that:

$$\forall (x, y, z) \in Q^3, \ \phi(x, y, z) = (1 - \lambda) \cdot f_1(x, y, z) + \lambda \cdot f_2(x, y, z) \qquad (2)$$

We will write $\phi = (f_1, f_2)[\lambda]$ to denote this relation. When $f_1 \neq f_2$ and $\lambda \in (0,1)$, we say that ϕ is a *diploid* ECA[2].

It can be noted that not all randomly-mixed ECAs are diploids and that the decomposition of a diploid is not always unique. The following proposition clarifies this fact.

---

[2] The name is composed from the ancient Greek διπλοῦς (diplous), which means twofold, double, and εἶδος (eidos), which evokes the form, the shape, the face, etc.

Proposition 1. *Let ϕ be a stochastic ECA, ϕ is a diploid if and only if there exists $\lambda \in (0,1)$ and $(x,y,z) \in Q^3$ such that $\phi(x,y,z) = \lambda$ and $\forall(x,y,z) \in Q^3, \phi(x,y,z) \in \{0, \lambda, 1 - \lambda, 1\}$.*

Indeed, if ϕ is a diploid such that $\phi = (f_1, f_2)[\lambda]$ then there exists a triplet $(x,y,z) \in Q^3$ such that $\phi(x,y,z) \notin \{0,1\}$, which implies that $\lambda = \phi(x,y,z)$ or $\lambda = 1 - \phi(x,y,z)$. Moreover, it can be noted that if the mixing rate λ is different from $1/2$, then the decomposition of a diploid into its components f_1 and f_2 is unique, up to the *exchange symmetry*, that is, the symmetry that exchanges f_1 and f_2 and inverts λ into $1 - \lambda$.

Let f_1 and f_2 be two different ECAs, we call a *diagonal* the set of diploids that can be obtained by combining f_1 and f_2 with a mixing rate $\lambda \in (0,1)$. Diagonals can be seen as lying on the diagonals of the hypercube $[0,1]^8$, where the vertices of the hypercube represent the deterministic ECAs. They are simply denoted by $(f_1, f_2) = \{(f_1, f_2)[\lambda], \lambda \in (0,1)\}$.

Since there are 256 ECAs, there exist $256 \times 256 = 65\,536$ couples of ECAs that allow one to define a randomly-mixed ECA. If one is interested in diagonals then there exist $256 \cdot (256 - 1)/2 = 32640$ couples to study, taking into account the exchange symmetry. However, the *conjugation symmetry*, which exchanges the 0s and 1s, and the *reflection symmetry*, which exchanges the left and right directions, can also be employed to reduce the number of couples that one may study to explore the space of diploid rules. Let us now enumerate the number of couples that are non-equivalent according to these symmetries.

Let R, C, T be the operators which respectively represent the reflection, the conjugation, and the composition of both operations. Formally for an ECA $f : Q^3 \to Q$, we define:

$$R(f) = f(z,y,x),$$
$$C(f) = 1 - f(1 - x, 1 - y, 1 - z),$$
$$T(f) = R \circ C(f) = C \circ R(f) = 1 - f(1 - z, 1 - y, 1 - x).$$

It is straightforward to generalise the definitions above to the stochastic rules, either by applying the symmetries to the local function or by applying them to the configurations that define the Markov chain of the ECA (see above).

Proposition 2. *Under the symmetries R, C, and T, there are 8808 non-equivalent diagonals that define the diploid rules.*

Proof. Let B be the set of ECAs which are invariant under both symmetries R and C (and thus T), let R, C, T be respectively the ECAs which are invariant under only the symmetry R, C, or T only and let N represent the rules which have no invariance. It can be easily verified that we have $|B| = 2^3 = 8$, $|C| = |T| = 2^4 - |B| = 8$, $|R| = 2^6 - |B| = 56$ and $|N| = 256 - (|B| + |R| + |C| + |T|) = 176$.

Let b, r, c, t, n represent the number of classes of equivalences of the sets B, R, C, T and N, respectively. We have $b = |B| = 8$, $r = |R|/2 = 28$, $c = |C|/2 = 4$, $t = |T|/2 = 4$ and $n = |N|/4 = 44$. The total number of non-equivalent ECAs is thus equal to $b + c + r + t + n = 88$, we represent each class of equivalence by the rule which has the smallest decimal code, the *minimal representative*.

We say that a couple of rules (f, g) is invariant under the symmetry S if (a): $(f, g) = (S(f), S(g))$ or (b): $(f, g) = (S(g), S(f))$. Unfortunately, it is not possible to obtain the equivalence classes of the diagonals by only considering the couples formed by the minimal representatives and the exchange symmetry. For example, if one considers the two ECAs 0 and 255, 0 is a minimal representative but not 255, but, obviously, it is not possible to form a couple of two minimal representative rules that would generate an equivalent diagonal.

Let $S = Q^{Q^3}$ denote the ECA space and $S_d = S \times S$ be the space of couples of ECAs. Taking similar notations as we did for the ECA space, let B_d be the set of couples that are invariant under both symmetries R and C, and R_d, C_d and T_d the set of couples that are invariant under *only* the symmetry R, C and T, respectively, and let N_d be the set of couples that have no invariance. Note that S_d contains 256 *repetitive couples*, that is, couples in the form (f, f).

Invariance by R and C. The set B_d is formed by the couples (f, g) such that: (1) $(f, g) = (R(f), R(g))$ or (2) $(f, g) = (R(g), R(f))$; and such that: (3) $(f, g) = (C(f), C(g))$ or (4) $(f, g) = (C(g), C(f))$. Let B_d^1 be the set of couples which verify (1) and (3), we have $|B_d^1| = |B \times B| = 8 \cdot 8 = 64$. Let B_d^2 be the set of couples which verify (1) and (4) and which do not belong to B_d^1. It can be easily verified that $B_d^2 = \{(f, g), f \in R, g = C(f)\}$, which gives $|B_d^2| = |R| = 56$. Let B_d^3 be the set of couples which verify (2) and (3) and which do not belong to B_d^1. Similarly, it can be easily verified that $B_d^3 = \{(f, g), f \in C, g = R(f)\}$, which gives $|B_d^3| = |C| = 8$. In a similar way: $B_d^4 = \{(f, g), f \in T, g = R(f) = C(f)\}$, which gives $|B_d^4| = |T| = 8$.

One may easily verify that the sets $B_d^1, B_d^2, B_d^3, B_d^4$ form a partition of B_d. We thus have: $|B_d| = 64 + 56 + 8 + 8 = 128 + 8 = 136$. Given that B_d contains 8 repetitive couples and given that each pair of non-repetitive couples (f, g) and (g, f) counts for one diagonal, the number of diagonals that can be generated by elements of B_d is $\#(B_d) = (136 - 8)/2 = 64$.

Invariance by R only. The set R_d is formed by the couples such that: (1) $(f, g) = (R(f), R(g))$ or (2) $(f, g) = (R(g), R(f))$, and such that: (3) $(f, g) \notin B_d$.

Let R_d^1 be the set of couples that verify condition (1) and not condition (2), and let R_d^2 be the set of couples that verify condition (2) but not condition (1). We have: $R_d^1 = (B \cup R) \times (B \cup R)$, which gives $|R_d^1| = 64 \cdot 64 = 4096$. We have: $R_d^2 = \{(f, g) \in S_d, f \in S \setminus (B \cup R), g = R(f)\}$, which gives: $|R_d^2| = 256 - (56 + 8) = 192$. As $R_d = (R_d^1 \cup R_d^2) \setminus B_d$, we find that $|R_d| = 4096 + 192 - 136 = 4152$. Given that R_d contains $|R| = 56$ repetitive couples, and given that each quadruplet (f, g), (g, f), $(C(f), C(g))$, $(C(g), C(f))$ counts for one non-equivalent diagonal, we have: $\#(R_d) = (4152 - 56)/4 = 4096/4 = 1024$.

Invariance by C only and by $R \circ C$ only. The same arguments as seen above can be repeated. We count: $|C_d| = |T_d| = 256 + 240 - 136 = 360$. The number of non-equivalent diagonals found in C_d and T_d is: $\#(T_d) = \#(C_d) = 88$.

No invariance. We now enumerate the number of couples which show no invariance by R, C or $R \circ C$. We have $N_d = S_d \setminus (B_d \cup R_d \cup C_d \cup T_d)$ and we count:

$|N_d| = 256.256 - (136 + 4152 + 360 + 360) = 65536 - 5008 = 60528$. Since this set of couples contains $|N| = 256 - (8 + 56 + 8 + 8) = 176$ repetitive couples and since the exchange symmetry and the symmetries C, R or $R \circ C$ allow one to associate 8 different couples in one equivalence class, we find that the number of non-equivalent diagonals that can be generated by elements of N_d is $\#(N_d) = (60528 - 176)/8 = 7544$.

We are now in position to add the set cardinals calculated above to find how many non-equivalent diagonals there are to generate all the possible diploids:
$\#(S_d) = 64 + 1024 + 88 + 88 + 7544 = 8808$.   □

To introduce an order on these 8808 couples of diploid generators, we may simply use the lexicographic order on the pair of decimal codes attached to each rule. In other words, given two pairs of rules (f, g) and (f', g') we say that $(f, g) \leq (f', g')$ if and only if $f < f'$ or if $f = f'$ and $g \leq g'$. As seen above, the couples may be grouped by sets of equivalent pairs of size 2 (elements of B_d), 4 (elements of R_d, C_d, T_d), or 8 (elements of N_d). We call a *minimal representative* the pair which has the smallest code in its class of equivalence.

3   First Steps in the Space of Diploids

As already mentioned, there exists a plethora of interesting questions that can be asked on stochastic cellular automata. As a first step, we propose to examine a simple question: given a diagonal (f_1, f_2), what kind of *qualitative change* may happen to the diploids of this diagonal when λ is varied? In particular, we want to know if there exists some values of λ where the behaviour of the system changes drastically. Here again, the answer we give may vary widely depending on how we estimate the "behaviour" of our cellular automaton. We thus simply propose to use two macroscopic parameters: the *density*, the ratio of the states 1 and the *kinks density*, the ratio of the occurrence of the length-2 patterns 01 and 10.

The exhaustive study of the 8808 diagonals is a daunting task, we thus need to focus on some families of diploids. We propose to begin our examination with the diagonals that are formed with the most simple rules: the null rule, the identity rule, and the inversion rule.

3.1   Experimental Protocol

For a given ring size n and a configuration $x \in Q^\mathcal{L}$, we call the *density* the ratio $d(x) = |x|_1/n$ where $|x|_1 = |\{i \in \mathcal{L}, x_i = 1\}|$ and *kinks density* the ratio $d_k(x) = (|x|_{10} + |x|_{01})/n$ where $|x|_{qq'} = |\{i \in \mathcal{L}, x_i = q \text{ and } x_{i+1} = q'\}|$. For each couple of rules (f, g) we will sample the value of d and d_k of the stochastic mixtures of two ECAs for a system composed of $n = 10\,000$ cells. We vary λ from 0 to 1 by increasing λ from 0 to 0.05 by steps of 0.01, then from 0.05 to 0.95 by steps of 0.05, and then from 0.95 to 1 by steps of 0.01. The step length is made smaller for high and small values of λ to give a more precise view close to

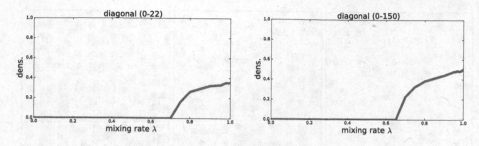

Fig. 1. Density as a function of λ for the diagonals (0,22) and (0,150).

the special points $\lambda = 0$ and $\lambda = 1$, where important modifications may occur. Indeed, the transition from a deterministic system to a non-deterministic one is likely to produce an abrupt change of behaviour; however, we leave for future work the study of such effects and we concentrate on the changes which occur for non-extremal values, as they are in some sense more surprising. For each diploid, we start from a uniform initial condition where each cell has an equal probability to be in state 0 or 1; we let the system evolve during $T_{\text{trans}} = 5\,000$ time steps and we sample the value of the parameters d and d_{k} to have a numerical estimation of the asymptotic density, that is, the average value we would obtain if the system's size would tend to infinity and if the sampling operation was repeated an infinite number of times. We sample only one value of this density for each value of α.

3.2 Mixtures with the Null Rule (ECA 0)

We applied the protocol described above for the 159 minimal non-equivalent diagonals in the form $(0, f)$. Figure 1 shows the estimation of the asymptotic density as function of the mixture rate in two diagonals where a qualitative modification of the behaviour occurs. This type of brutal change of behaviour is well-known and is called a second-order phase transition: the function $d(\lambda)$ is continuous but its derivative is not. There exists a critical mixing rate λ_c which separates a behaviour where the system converges to $0^{\mathcal{L}}$ (passive phase) and a behaviour with a stationary non-zero density (active phase). If the system were composed of an infinite number of cells, the property stated above would be exact. However, because of finite-size effects, it should be noted that even for the active phase, there exists a small but non-zero probability to reach the absorbing state $0^{\mathcal{L}}$, for example if all the cells apply rule 0 at the same time step. Regnault studied these type of phenomenon of convergence and he gave analytical results for a particular α-asynchronous ECA rule (see below) [10]. Second-order phase transitions were detected in no less than 34 diagonals $(0, f)$ and where f is found in the following ECA list: 18, 22, 26, 28, 30, 50, 54, 58, 60, 62, 78, 90, 94, 110, 122, 126, 146, 150, 154, 156, 158, 178, 182, 186, 188, 190, 202, 206, 218, 22, 234, 238, 250, 254 (see also Table 1).

It can be noted that the diagonals (0,250) and (0,254) correspond to two cases of *oriented site percolation*: in this problem an infinite grid is formed of

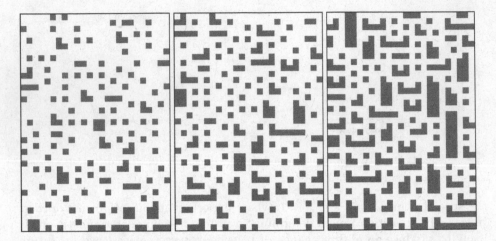

Fig. 2. Space-time diagrams of diploids (0,73) with (left): $\lambda = 0.25$, (middle): $\lambda = 0.50$, right: $\lambda = 0.75$. Time goes from bottom to top. Cells in state 0 and 1 are respectively represented by squares in white or blue (or dark gray). (Color figure online)

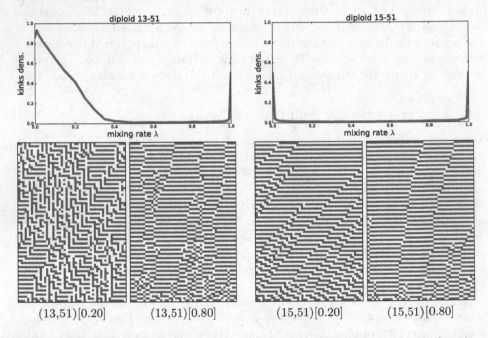

Fig. 3. Analysis of the diagonals (13,51) and (15,51). (top) Kinks density as a function of λ; (bottom) space-time diagrams for two arbitrary values of λ.

sites which can be open with probability p and closed with probability $1 - p$ and starting from a particular site, a fluid flows from open site to open site in a given direction. There exists a critical threshold for p for which the fluid

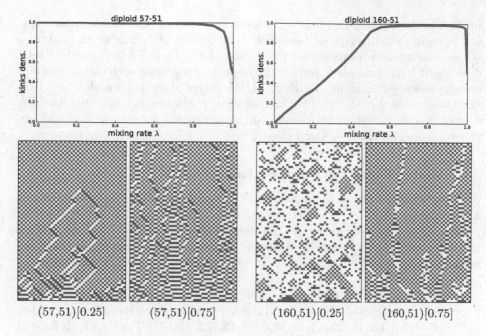

Fig. 4. Analysis of the diagonals (57,51) and (160,51). (top) Kinks density as a function of λ; (bottom) space-time diagrams for two arbitrary values of λ.

percolates, that is, has the possibility to travel arbitrary distances in the medium. The critical threshold has been measured with Monte Carlo simulations and various analytical estimates have been proposed. One may report to the study by Taggi [14] for more details and for a more general view on such *percolation operators* with various neighbourhoods.

We did not observe any phase transitions within the diagonals formed by ECA 0 and the ECAs with an odd code, that is, the rules for which $f(0,0,0) = 1$. Figure 2 shows the space-time diagrams of three diploids taken in the diagonal $(0,73)$ with $\lambda \in \{0.25, 0.5, 0.75\}$. These samples allow one to understand why the mixtures between rule 0 and an "odd" rule does not create any interesting structure: on the one hand, the null rule has a tendency to remove the 1s independently on any information on the neighbourhood. On the other hand, the odd rule may create a 1 in any place where there are three successive 0s. This idea needs of course to be studied with more attention and if possible by analytical means.

3.3 Mixtures with the Identity Rule: α-asynchronous ECA

The diploids defined with a mixture of the identity rule (ECA 204) and another ECA rule correspond to the α-asynchronous CA (where the mixing rate λ corresponds to the synchrony rate α). Such a family of rules has been studied quite in detail both numerically and analytically but there are still many challenges in their understanding [3,4,6,13]. The most surprising aspect of α-asynchronous CA is their great diversity of behaviour. There are 9 diagonals which contain a second-order phase transition for the density, but, contrarily to "mixed-with-0" diploids, it is not possible to predict the respective positions of the absorbing and active phases. For example, for ECA 50, the active phase is for high values of α and the absorbing phase is for low values, while the situation is inverted for ECA 6 or ECA 138. We refer to our survey on asynchronous for more information on this topic and on closely-related questions [6].

3.4   Mixtures with the Inversion Rule (ECA 51)

An interesting family of diploids to study is formed by the rules obtained as a mixture with ECA 51, that is, the rule $f(x,y,z) = 1 - y$. Contrarily to the previous families of diagonals, we did not observe any phase transition related to the evolution of the density. However, a new behaviour appears: there are some rules for which the *kinks* density take extremal values and become close (or equal) to 0 or 1.

First, let us examine the cases where the kinks density is close to 0. This happens in 16 diagonals $(f, 51)$ for the rules f which have the following codes: 1, 3, 5, 7, **9**, 11, **13**, 15, 25, 27, 29, 33, 37, **45**, **73**, and 77. The diagonals formed with rules written in bold font show the presence of a critical threshold: they converge to a zero kink value only for sufficiently high values of λ (see also Table 1). The other rules seem to remove the kinks for any value of the mixing rate. A peculiar behaviour appears for the diagonal (33,51) where the kinks disappear very slowly for high values of λ. Experiments do not show a critical phenomenon here. Figure 3 displays the evolution of the kinks density for the diagonals (13,51) and (15,51).

We represented examples of the space-time diagrams of the four diploids: (13,51)[.2],(13,51)[.8], (15,51)[.2], and (15,51)[.8] in Fig. 3. The space-time diagrams of the diploids which remove the kinks show that the qualitative behaviour of these rules is quite regular: they tend to form blinking regions of consecutive 0s and 1s. This blinking phenomenon is joined to a random walk of the frontiers. The regions disappear when their frontiers meet (annihilation phenomenon) which allows to infer that all these diploids solve the global synchronisation problem [7]: from any initial condition, they converge to the deterministic period-2 cycle where the uniform configurations $0^{\mathcal{L}}$ and $1^{\mathcal{L}}$ alternate. Interestingly, Richard has shown that it is not possible to solve this problem with a deterministic cellular automaton (see Ref. [11] and the extended version of Ref. [7]), but as it can be seen here, solving the problem with diploids is rather easy. It can be shown analytically that the convergence time of the diploid (15,51) scales quadratically with the number of cells. The proof of this property relies on a coupling with the shift rule (ECA 170) with an α-asynchronous updating [7].

Let us now turn our attention on the diagonals which show values of the kinks density which are close or equal to 1. There are 15 such diagonals $(f, 51)$ where f has the code: **32**, 34, **40**, 42, 56, 57, 58, **104**, **106**, **122**, **160**, 162, **168**, 170, **232**. The rules written in bold font show the presence of a critical

Table 1. List of diagonals where second-order phase transitions were observed. The first column shows the order parameter for which the critical phenomenon is observed, the second and third columns show the two elements of the diagonal.

Param.	Diagonal with	Rule list
Dens.	0 (null)	18, 22, 26, 28, 30, 50, 54, 58, 60, 62, 78, 90, 94, 110, 122, 126, 146, 150, 154, 156, 158, 178, 182, 186, 188, 190, 202, 206, 218, 22, 234, 238, 250, 254
Dens.	204 (identity)	6, 18, 26, 38, 50, 58, 106, 134, 146
Kinks dens.	51 (inversion)	9, 13, 32, 40, 45, 73, 104, 106, 122, 160, 168, 232

threshold: they converge to configurations where the states 0 and 1 alternate (high kinks density) only for a sufficiently high values of the mixing rate λ. The other diagonals seem to contain diploids which attain a kinks density of 1 for any value of $\lambda \in (0, 1)$. The diagonal (57,51) shows a slowdown of the convergence for high values of λ.

As it can be seen on Fig. 4, this behaviour can be associated to another form of symmetry breaking: the creation of a pattern where all the cells alternate their state in space and time. Here again, the synchronisation of the whole system is achieved by the random movements of the frontiers between non-synchronised regions, which act as defects which annihilate when they meet.

4   Discussion

This note constitutes a first step in the exploration of the space of the diploids. Such mixtures of two deterministic rules have been studied by various authors and it would be interesting to compile and "digest" the various phenomena that were studied on such models. Here, we systematically explored the mixtures formed with three "blind" rules which do not depend on their neighbourhood: the null rule, the identity rule and the inversion rule, and obtained some quite puzzling observations.

First, in the space of diploids, the occurrence of second-order phase transitions is not a rare phenomenon. So far, similar observations were linked to percolation phenomena [14], to the study of the effects of asynchronism [6], or rely on specific constructions. Note that if one takes a "randomly" chosen rule in the space of the elementary *stochastic rules* S_8, this rule is most likely a *positive-rate* rule, that is, for each neighbourhood there is a non-zero probability to reach both states. As a consequence, the chances to observe a phase transition by slightly modifying this rule is quite small (see Ref. [8] for more explanations). By contrast, the space of diploids may prove an appropriate subset of rules to find various forms of critical phenomena.

The second interesting aspect observed with our diploids is their ability to produce various forms of symmetry breaking: their inherent randomness allow them to easily create patterns with alternating states, or to reach homogeneous and synchronised blinking configurations. So far, diploids have provided simple solutions to the density classification problem (Fukś' rule, Schüle's rule and the traffic-majority rule) [5] or to the global synchronisation problem [7]. It is an open question to know how such models may perform some more complex forms of decentralised computations, and, if possible, solve problems that would be difficult or impossible to solve without the help of randomness[3].

References

1. Arrighi, P., Schabanel, N., Theyssier, G.: Stochastic cellular automata: correlations, decidability and simulations. Fundamenta Informaticae **126**(2–3), 121–156 (2013)
2. Bołt, W., Baetens, J.M., De Baets, B.: On the decomposition of stochastic cellular automata. J. Comput. Sci. **11**, 245–257 (2015). arXiv:1503.03318
3. Dennunzio, A., Formenti, E., Manzoni, L., Mauri, G.: m-asynchronous cellular automata: from fairness to quasi-fairness. Nat. Comput. **12**(4), 561–572 (2013)
4. Dennunzio, A., Formenti, E., Manzoni, L., Mauri, G., Porreca, A.E.: Computational complexity of finite asynchronous cellular automata. Theoret. Comput. Sci. **664**, 131–143 (2017)
5. Fatès, N.: Stochastic cellular automata solutions to the density classification problem - when randomness helps computing. Theory Comput. Syst. **53**(2), 223–242 (2013)
6. Fatès, N.: A guided tour of asynchronous cellular automata. J. Cell. Automata **9**(5–6), 387–416 (2014)
7. Fatès, N.: Remarks on the cellular automaton global synchronisation problem. In: Kari, J. (ed.) AUTOMATA 2015. LNCS, vol. 9099, pp. 113–126. Springer, Heidelberg (2015). doi:10.1007/978-3-662-47221-7_9
8. Mairesse, J., Marcovici, I.: Around probabilistic cellular automata. Theor. Comput. Sci. **559**, 42–72 (2014). Non-uniform Cellular Automata
9. Ricardo, J., Mendonça, G., de Mário, J., de Oliveira, M.J.: An extinction-survival-type phase transition in the probabilistic cellular automaton p182–q200. J. Phys. A Math. Theor. **44**(15), 155001 (2011)
10. Regnault, D.: Proof of a phase transition in probabilistic cellular automata. In: Béal, M.-P., Carton, O. (eds.) DLT 2013. LNCS, vol. 7907, pp. 433–444. Springer, Heidelberg (2013). doi:10.1007/978-3-642-38771-5_38
11. Richard, G.: On the synchronisation problem over cellular automata. To appear, Private communication (2017)
12. Schüle, M., Stoop, R.: A full computation-relevant topological dynamics classification of elementary cellular automata. Chaos **22**(4), 043143 (2012)
13. Silva, F., Correia, L., Christensen, A.L.: Modelling synchronisation in multi-robot systems with cellular automata: analysis of update methods and topology perturbations. In: Sirakoulis, G.C., Adamatzky, A. (eds.) Robots and Lattice Automata. ECC, vol. 13, pp. 267–293. Springer, Cham (2015). doi:10.1007/978-3-319-10924-4_12
14. Taggi, L.: Critical probabilities and convergence time of percolation probabilistic cellular automata. J. Stat. Phys. **159**(4), 853–892 (2015)
15. Turing, A.: The chemical basis of morphogenesis. Philos. Trans. Royal Soc. (London) **237**, 5–72 (1952)

---

[3] The author is grateful to Victor Roussanaly for his work on this topic as an intern in the Loria laboratory from June to August 2013. He also expresses his sincere recognition to Jordina Francès de Mas for her attention when reading the manuscript.

On the Computational Complexity
of the Freezing Non-strict Majority Automata

Eric Goles[1,2,3], Diego Maldonado[2(✉)], Pedro Montealegre[2],
and Nicolas Ollinger[2]

[1] Facultad de Ciencias y Tecnologia, Universidad Adolfo Ibañez, Santiago, Chile
[2] Univ. Orléans, LIFO EA 4022, 45067 Orléans, France
`diego.maldonado@uni-orleans.fr`
[3] LE STUDIUM, Loire Valley Institute for Advanced Studies, Orléans, France

Abstract. Consider a two dimensional lattice with the von Neumann neighborhood such that each site has a value belonging to $\{0,1\}$ which changes state following a freezing non-strict majority rule, i.e., sites at state 1 remain unchanged and those at 0 change iff two or more of it neighbors are at state 1. We study the complexity of the decision problem consisting in to decide whether an arbitrary site initially in state 0 will change to state 1. We show that the problem in the class **NC** proving a characterization of the maximal sets of stable sites as the tri-connected components.

1 Introduction

Majority automata can be defined as the two-state cellular automata, where in each synchronous step each site takes most represented state in its neighborhood. This kind of automata models many kinds of physic and social phenomena [1–5]. Let us denote by 1 and 0 the two states of the majority automata, which may represent respectively states *active* or *inactive*, *alive* or *dead*, etc. This paper is about the problem of predicting, given an initial configuration of a states, if a given site will change its state.

The computational complexity of a prediction problem can be defined as the amount of resources, like time or space, needed to predict it. In this case, we consider two fundamental classes: **P**, the class of problems solvable in polynomial time on a serial computer, and **NC**, the class of problems solvable in poly-logarithmic time in a PRAM machine, with a polynomial number of processors [6]. We say that **NC** is the class of problems which have a *fast parallel algorithm*. It is a well-known conjecture that $\mathbf{NC} \neq \mathbf{P}$, and so, if there exists "inherently sequential" problems, this is, problems that belong to **P** and do not belong to **NC**. The most likely to be inherently sequential are **P**-Complete problems, to which any other problem in **P** can be reduced (by an **NC**-reduction or a logarithmic space reduction). If any of these problems has a fast parallel algorithm, then $\mathbf{P} = \mathbf{NC}$ [6,7].

A. Dennunzio et al. (Eds.): AUTOMATA 2017, LNCS 10248, pp. 109–119, 2017.
DOI: 10.1007/978-3-319-58631-1_9

In [7] Moore studied the computational complexity of the majority automata in a d-dimensional lattice. He showed that in three or more dimensions the prediction problem is as hard as evaluating monotone circuits, which implies that the prediction problem is **P**-Complete. Roughly speaking, this means that in order to compute the state in a given site, the only option (unless **P** = **NC**) is to simulate the dynamics of the automaton until it reaches an attractor. However, Moore suggested that in two dimensions with von Neumann neighborhood it would be possible to predict exponentially faster, i.e., the problem is not **P**-Complete. In the same article, Moore also studied the *non-strict majority automata* (called also Half-or-More automata), which corresponds to the majority automata where the sites privilege state 1 over 0 in tie cases, not considering its own state in the neighborhood. Moore stated that the prediction problem for non-strict majority automaton is **P**-Complete in three or more dimensions, and also conjectured that in two dimensions the problem would not **P**-Complete.

In this article we study the prediction problem on the *freezing non-strict majority automata*. The freezing property means that a site in state 1 remains in that state in every future time step. Freezing automata model forest fires [8], infection spreading [9], bootstrap percolation [10] and voting systems [7]. Theoretical facts about those automata can be seen in [11].

The prediction problem for the *freezing majority automata* was studied by Goles et al. in [12], where the authors show that the prediction problem for the freezing majority automata is in **NC**, restricted to a two dimensional lattice with von Neumann neighborhood. This result is based on a characterization of *stable* sets of sites. A set of sites is called stable if a site in 0 remain in state 0 on any future time-step. The authors showed that for the freezing majority automata the stable sets can be characterized in terms of connected and biconnected components of the sets of sites initially in state 0. The prediction algorithm uses fast parallel algorithms computing connected and biconnected components due to Jájá [13].

In this paper, we show that the prediction problem for the non-strict majority automata is in **NC**. Our algorithm is based in a characterization of the stable sets for this rule. Unfortunately, the characterizations of stable sets for the freezing majority automaton is not valid for the non-strict one. In its place, we show that the stable sets in this case are roughly a set of sites initially in state 0 which form a tri-connected component, i.e., there are three disjoint paths between every pair of sites in the set. Then, we use a fast-parallel algorithm due to Jájá [14] to compute the tri-connected components of the sites initially in 0.

The article is organized as follows. In next section, we begin with the main formal definitions. In Sect. 3 we present a characterization of the stable sets for the freezing non-strict majority automata. In Sect. 4 we use the characterization of the previous section to obtain the main result. Finally, in Sect. 5 we conclude this article with a discussion and some open questions.

2 Preliminaries

Let us consider the freezing non-strict majority cellular automata (FNSMCA) as the cellular automata defined by the tuple $(\mathbb{Z}^2, \{0,1\}, N, f)$, where

$N = \{(0,0),(1,0),(-1,0),(0,1),(0,-1)\}$ is the von Neumann neighborhood and $f : \{0,1\}^5 \to \{0,1\}$ the local freezing non-strict majority function:

$$f(x) = \begin{cases} 1 \text{ if } x_1 = 1 \\[2mm] 1 \text{ if } (x_1 = 0) \wedge \left(\sum_{i \in [5]} x_i \geq 2 \right) \\[2mm] 0 \text{ otherwise} \end{cases}$$

For $c \in \{0,1\}^{\mathbb{Z}^2}$ and $i,j \in \mathbb{Z}$, call $c_{(i,j)+N}$ the vector $((i,j),(i+1,j),(i-1,j),(i,j+1),(i,j-1))$. The new state of each site of the lattice is computed synchronously, i.e., every site is updated at the same time, which is equivalent to the application of the global function $F : \{0,1\}^{\mathbb{Z}^2} \to \{0,1\}^{\mathbb{Z}^2}$ with $F(c)_v = f(c_{v+N})$.

The following definition characterizes the sites that are initially in 0 and never change, we call such sites *stable*.

Definition 1. *Given a configuration $c \in \{0,1\}^{\mathbb{Z}^2}$, we say that a site v is* stable *if and only if $c_v = 0$ and it remains at state 0 after any iterated application of the global rule, i.e., $F^t(c)_v = 0$ for all $t \geq 0$.*

One property of the strict majority automata that will be useful in our proofs is the *monotonicity*. For two configurations c and c', denote by $\leq$ the partial order relation over configurations, where $c \leq c'$ if and only if $c_u \leq c'_u$ for every $u \in \mathbb{Z}^2$. A function $G : \{0,1\}^{\mathbb{Z}^2} \to \{0,1\}^{\mathbb{Z}^2}$ is called *monotone* if $c \leq c'$ implies that $G(c) \leq G(c')$. Clearly, the strict majority automata (and its freezing version) is monotone.

A configuration $c \in \{0,1\}^{\mathbb{Z}^2}$ is called a *periodic configuration* if c consists in the periodic repetition of a finite configuration $x \in \{0,1\}^{[n] \times [m]}$, for $n, m > 0$. In such a case, we also call $c = c(x)$. Note that a FNSMCA in a periodic configuration $c(x)$, with $x \in \{0,1\}^{[n] \times [n]}$, reaches a fixed point in at most n^2 steps. Indeed, at each step before the dynamic reaches the fixed point, at least one site change from state 0 to state 1.

A natural problem consists in computing, given a periodic configuration c, the configuration obtained by the automata when it reaches the fixed point corresponding to c. We define the decision version of this problem, called (PREDICTION), as the problem consisting in decide, given a configuration, if a specific site initially in state 0 is not stable.

Prediction (PREDICTION)
Input: A finite configuration x of dimensions $n \times n$ and a site $u \in [n] \times [n]$ such that $x_u = 0$.
Question: Does there exists $T > 0$ such that $F^T(c(x))_u = 1$?

Clearly, if this problem is in **NC**, then we can compute the fixed point of the automaton running $\mathcal{O}(n^2)$ copies of the algorithm in each site initially in state 0.

Indeed, the fixed point obtained from c can be computed choosing state 1 on all sites that are not stable for configuration c.

For a finite set of sites $S \subseteq \mathbb{Z}^2$, we call $G[S] = (S, E)$ the graph defined with vertex set S, where two vertices are adjacent if the corresponding sites are neighbors for the von Neumann neighborhood. For a graph $G = (V, E)$, a sequence of vertices $P = v_1, \ldots, v_k$ is called a v_1, v_k- path if $\{v_i, v_{i+1}\}$ is an edge of G, for each $i \in [k]$. Two u, v-paths P_1, P_2 are called *disjoint* if $P_1 \cap P_2 = \{u, v\}$.

Definition 2. *A graph G is called tri-connected if for every pair of vertices $u, v \in V(G)$, G contains three disjoint u, v-paths.*

A maximal set of vertices of a graph G that induces a tri-connected subgraph is called a *tri-connected component* of G. The following proposition states that is decidable in **NC** if a graph is tri-connected, moreover, gives a fast-parallel algorithm computing the tri-connected components of an input graph.

Proposition 1 ([14]). *There is an algorithm that computes the tri-connected components of a graph in time $\mathcal{O}(\log^2 n)$ with $\mathcal{O}(n^4)$ processors.*

Consider the relation R over vertices of a graph, which states that two vertices u and v are related by R if there exist three disjoint paths connecting u and v. For a pair of vertices u and v of a graph G, call $G_{u,v}$ the transitive closure of graph $G - \{u, v\}$. The transitive closure of a graph G is the graph in the same vertex set, where each connected component is a clique. In [14] it is shown that two vertices s, t are related by R if they are connected in $G_{u,v}$ for every pair $u, v \in V(G) \backslash \{s, t\}$. Moreover, the transitive closure can be computed in time $\mathcal{O}(\log^2 n)$ with $\mathcal{O}(n^2)$ processors [14]. The tri-connected components are then constructed roughly as follows: for every triple u, v, w of vertices of G that are mutually tri-connected (i.e. related by R), the set $\{l \in V : lRu \wedge lRv \wedge lRw\}$ is the tri-connected component that contains vertices u, v and w.

3   A Characterization of Stable Sets

In this section, we characterize *stable sets* of a configuration c, i.e. sets of sites that are stable for c.

Lemma 1. *Let $x \in \{0, 1\}^{n^2}$ be a finite configuration and $u \in [n] \times [n]$ a site. Then, u is stable for $c = c(x)$ if and only if there exist a set $S \subseteq [n] \times [n]$ such that:*

- *$u \in S$,*
- *$c_u = 0$ for every $u \in S$, and*
- *$G[S]$ is a graph of minimum degree 3.*

Proof. Suppose that u is stable and let S be the subset of $[n] \times [n]$ containing all the sites that are stable for c. We claim that S satisfy the desired properties. Indeed, since S contains all the sites stable for c, then u is contained in S. On the other hand, since the automata is freezing, all the sites in S must be in state 0 on the configuration c. Finally, if $G[S]$ contains a vertex v of degree less than 3, it means that necessarily the corresponding site v has two non-stable neighbors that become 1 in the fixed point reached from c, contradicting the fact that v is stable.

On the other direction suppose that S contains a site that is not stable and let $t > 0$ be the minimum step such that a site v in S changes to state 1, i.e., $v \in S$ and t are such $F^{t-1}(c)_w = 0$ for every $w \in S$, and $F^t(c)_v = 1$. This implies that v has at least two neighbors in state 1 in the configuration $F^{t-1}(c)$. This contradicts the fact that v has three neighbors in S. We conclude that all the sites contained in S are stable, in particular u. □

For a finite configuration $x \in \{0,1\}^{[n]^2}$, let $D(x) \in \{0,1\}^{\{-n^2-n,\dots,n^2+2n\}^2}$ be the finite configuration of dimensions $m \times m$, where $m = 2n^2 + 3n$, constructed with repetitions of configuration x in a rectangular shape, as is depicted in Fig. 1, and sites in state 0 elsewhere. We also call $D(c)$ the periodic configuration $c(D(x))$. It is important to distinguish between $c(x)$ and $c(D(x))$: the first one is the periodic configuration defined as the repetition of the finite configuration x, while $c(D(x))$ corresponds to the periodic configuration obtained as repetitions of $D(x)$.

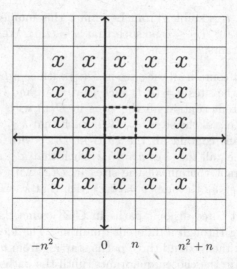

Fig. 1. Construction of the finite configuration $D(x)$ obtained from a finite configuration x of dimension $n \times n = 2 \times 2$. Note that $D(x)$ is of dimensions 14×14.

Lemma 2. *Let $x \in \{0,1\}^{[n]^2}$ be a finite configuration, and let u be a site in $[n] \times [n]$ such that $x_u = 0$. Then u is stable for $c = c(x)$ if and only if it is stable for $D(c)$.*

Proof. Suppose first that u is stable for c, i.e. in the fixed point c' reached from c, $c'_u = 0$. Call c'' the fixed point reached from $D(c)$. Note that $D(c) \leq c$ (where $\leq$ represent the inequalities coordinate by coordinate). Since the FNSMCA automata is monotonic, we have that $c'' \leq c'$, so $c''_u = 0$. Then u is stable for $D(c)$.

Conversely, suppose that $u \in [n] \times [n]$ is not stable for c, and let S be the set of all sites at distance at most n^2 from u. We know that in each step on the dynamics of c, at least one site in the periodic configuration changes its state, then in at most n^2 steps the site u will be in state 1. In other words, the state of u depends only on the states of the sites at distance at most n^2 from u. Note that for every $v \in S$, $c_v = D(c)_v$. Therefore, u is not stable in $D(c)$.   □

The set of sites (i,j) of $D(x)$ satisfying $(i,j) \in [m] \times ([-n^2 - n, -n^2 - 1] \cup (n^2 + n + 1, n^2 + 2n])$ or $(i,j) \in ([-n^2 - n, -n^2 - 1] \cup (n^2 + n + 1, n^2 + 2n]) \times [m]$, are called the *border B* of $D(x)$. Note that the border of $D(x)$ contains only sites in state 0. We call $D(x) - B$ the *interior* of $D(x)$. Note that B is tri-connected and forms a set of sites stable for $D(c)$ thanks to Lemma 1. We call Z the set of sites w in $[m] \times [m]$ such that $D(x)_w = 0$.

Lemma 3. *Let u be a site in $[n] \times [n]$ stable for $D(c)$. Then, there exist three disjoint paths on $G[Z]$ connecting u with sites of the border B. Moreover, the paths contain only sites that are stable for $B(c)$.*

Proof. Suppose that u is stable. From Lemma 1 this implies that u has three stable neighbors. Let $0 \leq i, j \leq n$ be such that $u = (i,j)$. We divide the interior of $D(c)$ in four quadrants:

- The first quadrant contain all the sites in $D(x)$ with coordinates at the north-east of u, i.e., all the sites $v = (k,l)$ such that $k \geq i$ and $l \geq j$.
- The second quadrant contain all the sites in $D(x)$ with coordinates at the north-west of u, i.e., all the sites $v = (k,l)$ such that $k \leq i$ and $l \geq j$.
- The third quadrant contain all the sites in $D(x)$ with coordinates at the south-west of u, i.e., all the sites $v = (k,l)$ such that $k \leq i$ and $l \leq j$.
- The fourth quadrant contain all the sites in $D(x)$ with coordinates at the south-east of u, i.e., all the sites $v = (k,l)$ such that $k \geq i$ and $l \leq j$.

We will construct three disjoint paths in $G[Z]$ connecting u with the border, each one passing through a different quadrant. The idea is to first choose three quadrants, and then extend three paths starting from u iteratively picking different stable sites in the chosen quadrants, until the paths reach the border.

Suppose without loss of generality that we choose the first, second and third quadrants, and let u_1, u_2 and u_3 be three stable neighbors of u, named according to Fig. 2.

Starting from u, u_1, we extend the path P_1 through the endpoint different than u, picking iteratively a stable site at the east, or at the north if the site in the north is not stable. Such sites will always exist since by construction the current endpoint of the path will be a stable site, and stable sites must have three stable neighbors (so either one neighbor at east or one neighbor at north).

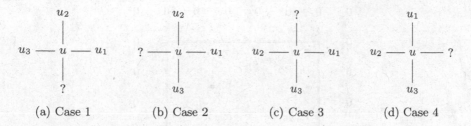

| (a) Case 1 | (b) Case 2 | (c) Case 3 | (d) Case 4 |

Fig. 2. Four possible cases for u_1, u_2 and u_3. Note that one of these four cases must exist, since u has at least three stable neighbors. From u_1 we will extend a path through the first quadrant, from u_2 a path through the second quadrant, and from u_3 a path through the third quadrant.

The iterative process finishes when P_1 reaches the border. Note that necessarily P_1 is contained in the first quadrant. Analogously, we define paths P_2 and P_3, starting from u_2 and u_3, respectively, and extending the corresponding paths picking neighbors at the north-west or south-west, respectively. We obtain that P_2 and P_3 belong to the second and third quadrants, and are disjoint from P_1 and from each other.

This argument is analogous for any choice of three quadrants. We conclude there exist three disjoint paths of stable sites from u to the border B. □

Lemma 4. *Let u, v be two sites in $[n] \times [n]$ stable for $D(c)$. Then, there exist three disjoint u, v-paths in $G[Z]$ consisting only of sites that are stable for $D(c)$.*

Proof. Let u, v be stable vertices. Without loss of generality, we can suppose that $u = (i, j)$, $v = (k, l)$ with $i \leq k$ and $j \leq l$ (otherwise we can rotate x to obtain this property). In this case u and v divide the interior of $D(x)$ into nine regions (see Fig. 3). Let $P_{u,2}, P_{u,3}, P_{u,4}$ be three disjoint paths that connect u with the border through the second, third and fourth quadrants of u. These paths exist according to the proof of Lemma 3. Similarly, define $P_{v,1}, P_{v,2}, P_{v,3}$ three disjoint paths that connect v to the border through the first, second and third quadrants of v.

Observe fist that $P_{u,3}$ touches regions that are disjoint from the ones touched by $P_{v,1}, P_{v,2}$ and $P_{v,3}$. The same is true for $P_{v,1}$ with respect to $P_{u,2}, P_{u,3}, P_{u,4}$. The first observation implies that paths $P_{u,3}$ and $P_{v,1}$ reach the border without intersecting any other path. Let w_1 and w_2 be respectively the intersections of $P_{u,3}$ and $P_{v,1}$ with the border. Let now P_{w_1, w_2} be any path in G_B connecting w_1 and w_2. We call $P_{1,3}$ the path induced by $P_{u,3} \cup P_{w_1,w_2} \cup P_{v,1}$.

Observe now that $P_{u,2}$ and $P_{v,4}$ must be disjoint, as well as $P_{u,4}$ and $P_{v,2}$. This observation implies that $P_{u,2}$ either intersects $P_{v,2}$ or it do not intersect any other path, and the same is true for $P_{u,4}$ and $P_{v,4}$. If $P_{u,2}$ does not intersect $P_{v,2}$, then we define a path $P_{2,2}$ in a similar way than $P_{1,3}$, i.e., we connect the endpoints of $P_{u,2}$ and $P_{v,2}$ through a path in the border (we can choose this path disjoint from $P_{1,3}$ since the border is tri-connected). Suppose now that $P_{u,2}$ intersects $P_{v,2}$. Let w the first site where $P_{u,2}$ and $P_{v,2}$ intersect, let $P_{u,w}$ be the

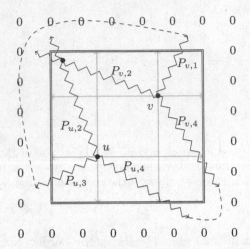

Fig. 3. Vertices u and v divides the interior of $D(x)$ into four regions each. Together they split the space into nine regions. According to Lemma 3), we can choose three disjoint paths connecting u and v, in such a way that each of the nine regions intersect at most one path. We use the border of $D(x)$ to connect the paths that do do not intersect in the interior of $D(x)$.

u, w-path contained in $P_{u,2}$, and let $P_{w,v}$ be the w, v-path contained in $P_{v,2}$. We call in this case $P_{2,2}$ the path $P_{u,w} \cup P_{w,v}$. Note that also in this case $P_{2,2}$ is disjoint from $P_{1,3}$. Finally, we define $P_{4,4}$ in a similar way using paths $P_{u,4}$ and $P_{v,4}$. We conclude that $P_{1,3}$, $P_{2,2}$, and $P_{4,4}$ are three disjoint paths of stable sites connecting u and v in $G[Z]$. $\qquad\square$

We are now ready to show our characterization of stable set of vertices.

Theorem 1. *Let* $x \in \{0,1\}^{[n]^2}$ *be a finite configuration, and let* u *be a site in* $[n] \times [n]$. *Then,* u *is stable for* $c = c(x)$ *if and only if* u *is contained in a tri-connected component of* $G[Z]$.

Proof. From Lemma 2, we know that u is stable for c if and only if it is stable $D(c)$. Let S be the set of sites stable for $D(c)$. We claim that S is a tri-connected component of $G[Z]$. From Lemma 4, we know that for every pair of sites in S there exist three disjoint paths in $G[S]$ connecting them, so the set S must be contained in some tri-connected component T of $G[Z]$. Since $G[T]$ is a graph of degree at least three, and the sites in T are contained in Z, then Lemma 1 implies that T must form a stable set of vertices, then T equals S.

On the other direction, Lemma 1 implies that any tri-connected component of $G[Z]$ must form a stable set of vertices for $D(c)$, so u is stable for c. $\qquad\square$

4   The Algorithm

We are now ready to show the main result of this paper.

Theorem 2. PREDICTION *is in* **NC**.

Proof. Let (x, u) be an input of PREDICTION, i.e. x is a finite configuration of dimensions $n \times n$, and u is a site in $[n] \times [n]$. Our algorithm for PREDICTION first computes from x the finite configuration $D(x)$. Then, the algorithm uses the algorithm of Proposition 1 to compute the tri-connected components of $G[Z]$, where Z is the set of sites w such that $D(x)_w = 0$. Finally, the algorithm answers *no* if u belongs to some tri-connected component of $G[Z]$, and answer *yes* otherwise.

Algorithm 1. PREDICTION

Input: x a finite configuration of dimensions $n \times n$ and $u \in [n] \times [n]$ such that $x_u = 0$.

1: Compute the finite configuration $D(x)$ of dimensions $m \times m$ with $m = 2n^2 + 3n$
2: Compute the set $Z = \{w \in [m] \times [m] \ : \ D(x)_w = 0\}$.
3: Compute the graph $G[Z]$.
4: Compute the set $\mathcal{T}$ of tri-connected components of $G[Z]$.
5: **for** each $T \in \mathcal{T}$ **do**
6:   **if** $u \in T$ **then**
7:     **return** *no*
8:   **end if**
9: **end for**
10: **return** *yes*

The correctness of Algorithm 1 is given by Theorem 1. Indeed, the algorithm answers *yes* on input (x, u) only when u does not belong to a tri-connected component of $G[Z]$. From Theorem 1, it means that u is not stable, so there exists $t > 0$ such that $F^t(c(x))_u = 1$.

Step **1** can be done in $\mathcal{O}(\log n)$ time with $m^2 = \mathcal{O}(n^6)$ processors: one processor for each site of $D(x)$ computes from x the value of the corresponding site in $D(x)$. Step **2** can be done in time in $\mathcal{O}(\log m) = \mathcal{O}(\log n)$ with $\mathcal{O}(m^2)$ processors, representing Z as a vector in $\{0, 1\}^{m^2}$, each coordinate is computed by a processor. Step **3** can be done in time $\mathcal{O}(\log n)$ and $\mathcal{O}(m^2)$ processors: we give one processor to each site in Z, which fill the corresponding four coordinates of the adjacency matrix of $G[Z]$. Step **4** can be done in time $\mathcal{O}(\log^2 n)$ with $\mathcal{O}(n)$ processors using the algorithm of Proposition 1. Finally, steps **5** to **10** can be done in time $\mathcal{O}(\log n)$ with $\mathcal{O}(n^2)$ processors: the algorithm checks in parallel if u is contained in each tri-connected components. There are $\mathcal{O}(n)$ tri-connected components, each of them containing $\mathcal{O}(n)$ elements. All together the algorithm runs in time $\mathcal{O}(\log^2 n)$ with $\mathcal{O}(n^6)$ processors.

5   Conclusion

We showed that the prediction problem for the two-dimensional freezing non-strict majority automaton is in **NC**. This question was posed in [7,12].

In [12] it is shown that the prediction problem for freezing non-strict majority automaton on an arbitrary graph of degree at most four is **P**-Complete, and in graphs of degree at most three is in **NC**. The authors conjectured that this problem is in **NC** on any planar topology. We remark that if we remove the hypothesis that the topology is a grid, then our characterization of stable sets (Theorem 1) is no longer true, even for planar regular graphs of degree four. Indeed a regular graph of degree four might not be tri-connected (see Fig. 4).

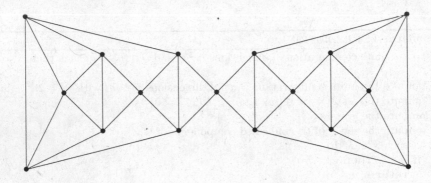

Fig. 4. Non-triconnected degree 4 graph. If all the vertices are initially in state 0, then the configuration is stable.

In our prediction problem, our goal was to compute the configuration obtained once the fixed point is reached. A variant of this problem could consider the complexity of the problem consisting in, given a configuration c and $T > 0$ and a site v, compute state of v in the configuration obtained after T steps, i.e., compute $F^T(c)_v$. Our algorithm is not valid for this problem, since a site might not be stable for the input configuration, but change its state in more than T steps. We believe that this version of the prediction problem is harder than the one treated in this paper. Indeed, we can create very simple gadgets that allow to simulate planar monotone circuitry only for a fixed number of steps, but that are destroyed when we do not bound the time of the simulation. To our knowledge, there are no example of an automata network capable to both simulate planar monotone circuitry, and which corresponding prediction problem is in **NC**.

References

1. Schelling, T.C.: Micromotives and Macrobehavior. WW Norton & Company, New York (2006)
2. Medina, P., Goles, E., Zarama, R., Rica, S.: Self-organized societies: on the Sakoda model of social interactions. Complexity **2017** (2017)
3. Domic, N.G., Goles, E., Rica, S.: Dynamics and complexity of the schelling segregation model. Phys. Rev. E **83**(5), 056111 (2011)

4. Castellano, C., Fortunato, S., Loreto, V.: Statistical physics of social dynamics. Rev. Modern Phys. **81**(2), 591 (2009)
5. Hegselmann, R.: Modeling social dynamics by cellular automata. In: Computer Modeling of Social Processes, pp. 37–64 (1998)
6. Greenlaw, R., Hoover, H., Ruzzo, W.: Limits to Parallel Computation: P-Completeness Theory. Oxford University Press Inc., New York (1995)
7. Moore, C.: Majority-vote cellular automata, ising dynamics, and p-completeness. Working papers, Santa Fe Institute (1996)
8. Karafyllidis, I., Thanailakis, A.: A model for predicting forest fire spreading using cellular automata. Ecol. Model. **99**(1), 87–97 (1997)
9. Fuentes, M.A., Kuperman, M.N.: Cellular automata and epidemiological models with spatial dependence. Physica A: Stat. Mech. Appl. **267**(3), 471–486 (1999)
10. Chalupa, J., Leath, P.L., Reich, G.R.: Bootstrap percolation on a bethe lattice. J. Phys. C: Solid State Phys. **12**(1), L31 (1979)
11. Goles, E., Ollinger, N., Theyssier, G.: Introducing freezing cellular automata. In: 21st International Workshop (AUTOMATA 2015) Cellular Automata and Discrete Complex Systems, Turku, Finland, vol. 24. TUCS Lecture Notes, pp. 65–73, June 2015
12. Goles, E., Montealegre-Barba, P., Todinca, I.: The complexity of the bootstraping percolation and other problems. Theor. Comput. Sci. **504**, 73–82 (2013)
13. JáJá, J.: An Introduction to Parallel Algorithms. Addison Wesley Longman Publishing Co., Inc., Redwood City (1992)
14. JáJá, J., Simon, J.: Parallel algorithms in graph theory: planarity testing. SIAM J. Comput. **11**(2), 314–328 (1982)

Distortion in One-Head Machines and Cellular Automata

Pierre Guillon[1] and Ville Salo[2(⊠)]

[1] Université d'Aix-Marseille CNRS,
Centrale Marseille, I2M, UMR 7373, 13453 Marseille, France
pierre.guillon@math.cnrs.fr
[2] University of Turku, Turku, Finland
vosalo@utu.fi

Abstract. We give two families of examples of automorphisms of subshifts that are *range-distorted*, that is, the radius of their iterations grows sublinearly. One of these families comes from one-head machines, and allows us to build such automorphisms for the full shift, and to obtain undecidability results. We also give some conditions on the functions that can occur as such growths.

1 Introduction

In this article, 'distortion' means that something that typically grows or moves linearly or not at all instead does so at an intermediate rate. In one-head machines, we consider sublinear head movement (the head visits $o(t)$ cells in t steps), and in cellular automata sublinear radius growth (the radius of the iterates grows in $o(t)$), which corresponds to range distortion in the terminology of [1]. In both cases, we show 'trichotomy' results: there are logarithmic gaps between periodic and distorted cases, and between distorted and positive-speed machines.

We show that every aperiodic one-head machine is distorted. The existence of aperiodic one-head machines is well-established, in particular [2] shows that they not only exist but form a computationally hard (undecidable) set. The single most beautiful example of an aperiodic machine is probably the SMART machine [3], whose moving tape dynamics is even minimal.

We discuss two ways of achieving distortion in automorphism groups of subshifts. To every one-head machine, we can associate a cellular automaton (on a full shift) whose radius grows at roughly the same speed as the head of the one-head machine moves. Given that there exist distorted one-head machines, there also exist distorted cellular automata. The examples given are reversible, and thus we obtain distorted automorphisms on a transitive subshift, answering an implicit question of [1]. By known embedding theorems, we obtain such examples on all uncountable sofic shifts.

A. Dennunzio et al. (Eds.): AUTOMATA 2017, LNCS 10248, pp. 120–138, 2017.
DOI: 10.1007/978-3-319-58631-1_10

We also construct an example of an automorphism on a general subshift with 'highly unbalanced distortion', in the sense that for an infinite set of times $t \in \mathbb{N}$, f^t has a 'right-leaning' neighborhood (one of the form $[\![a, \infty[\![)$ that contains only slightly more than logarithmically many cells to the left of the origin, and 'left-leaning' neighborhood with the symmetric property, yet all its two-sided neighborhoods grow at an almost linear rate. In particular, the intersection of all neighborhoods is far from being a neighborhood, answering Question 3.26 of [1].

2 Definitions

2.1 Subshifts and Cellular Automata

Let Σ be a finite set called the *alphabet*. Then $\Sigma^{\mathbb{Z}}$ with the product topology is called the *full shift*, and it is a $\mathbb{Z}$-dynamical system under the shift map $\sigma : \Sigma^{\mathbb{Z}} \to \Sigma^{\mathbb{Z}}$ defined by $\sigma(x)_i = x_{i+1}$. Closed shift-invariant subsets of it are called *subshifts*.

If X and Y are subshifts, a function $f : X \to Y$ is called a *morphism* if it is continuous and $\sigma \circ f = f \circ \sigma$. It is an *endomorphism* if $Y = X$ and an *automorphism* if, besides, it is bijective (in which case it automatically has a left and right inverse endomorphism). A *cellular automaton* is another name for an endomorphism, though often this term is reserved for the case $X = Y = \Sigma^{\mathbb{Z}}$. Automorphisms are also called *reversible cellular automata*.

An endomorphism f is *preperiodic* if $f^{p+q} = f^q$ for some preperiod $q \in \mathbb{N}$ and some period $p \in \mathbb{N} \backslash \{0\}$. If an automorphism is preperiodic with $q = 0$, it is *periodic*.

The trace map $T_f : X \to \Sigma^{\mathbb{N}}$ is the map defined by $T_f(x)_t = f^t(x)_0$ for all $x \in X$ and $t \in \mathbb{N}$. It is clear that $\tau_f = T_f(X)$ is a one-sided subshift (closed and shift-invariant), which is finite if and only if f is preperiodic.

For X a subshift and $n \in \mathbb{N}$ we define the *complexity* function $\mathcal{K}_X$ by $\mathcal{K}_X(n) = \left| \left\{ x_{[\![0,n[\![} \mid x \in X \right\} \right|$, the number of distinct patterns occuring in configurations of X. It is easy to see that if X is infinite, then $\mathcal{K}_X$ is increasing.

2.2 Neighborhoods and Radii

It is quite well-known [4] that if $f : X \to Y$ is a morphism, then it admits a neighborhood, that is a finite interval $I \subset \mathbb{Z}$ such that $\forall x, y \in X, x_I = y_I \Rightarrow f(x)_0 = f(y)_0$.

Let X and Y be $\mathbb{Z}$-subshifts and $f : X \to Y$ a morphism. We define the set of *neighborhoods* as

$$N(f) = I = [\![a, b]\!] \subset \mathbb{Z} \mid \forall x, y \in X, x_I = y_I \implies f(x)_0 = f(y)_0.$$

The *diameter* $D(f)$ of a morphism f is then the least possible diameter $2r+1$ of a central neighborhood $[\![-r, r]\!] \in N(f)$.

Remark 1. It is easy to see that $N(f)$ is an upset: $I \in N(f), J \supset I \implies J \in N(f)$.

The case when $N(f)$ is a principal filter, that is when $N(f) = \{J \mid J \supset I\}$ for some finite interval I (it is well-known that this happens for the full shift), is especially desirable. In that case, I must be the intersection of all elements in $N(f)$, and thus we define $I(f) = \bigcap_{J \in N(f)} J$ (it corresponds to $\mathcal{I}(-1, f)$ in the notation of [1]). Let us also define $d(f)$ as the diameter of $\{0\} \cup I(f)$, which is at most $D(f)$ (and is equal if X is the full shift for example). Theorem 5 will give an example of endomorphism where $N(f)$ is far from being a filter.

2.3 Distortion

Let f be an endomorphism of a subshift X. For $t \in \mathbb{N}$, let us define $D_t(f)$ as $\max_{k \leq t} D(f^k)$. It is clear that $D_t(f) \leq t(D(f) - 1) + 1 = O(t)$, and that $D_t(f) \leq D_{p+q}(f)$ is bounded if $f^{p+q} = f^q$. f will be called *range-distorted* (or simply, in this article, *distorted*) if $D_t(f) = o(t)$ but f is not preperiodic.

This definition is equivalent to the one from [1,5], and comparable to the notion of distortion from group theory: if f^t can be expressed as a product of $o(t)$ generators of some finitely generated endomorphism submonoid, then f is range-distorted.

The distortion function $t \mapsto D_t(f)$ cannot be arbitrarily low. In fact, naively counting the possible local rules gives a $\log \log$ lower bound, but the Morse-Hedlund theorem allows to 'remove' one log in the following proposition, which is a direct adaptation from the main argument in [5, Theorem 3.8].

Proposition 1. *If X is a subshift and $f : X \to X$ an endomorphism. Then exactly one of the following holds:*

- *(D_t) is bounded (f is preperiodic);*
- *$\forall t \in \mathbb{N}, D_t(f) \geq \mathcal{K}_X^{-1}(\mathcal{K}_{\tau_f}(t)) > \mathcal{K}_X^{-1}(t) = \Omega(\log t)$ and $D_t(f) = o(t)$ (f is distorted);*
- *$D_t(f) = \Theta(t)$, and $d(f^t) = \Theta(t)$ (f has non-0 Lyapunov exponents).*

Note that if X has linear complexity or if f has positive entropy, then the central class is empty. Moreover, if $\mathcal{K}_X(n) = O(n^d)$ (resp. $O(2^{n^\varepsilon})$), then endomorphisms of this class must even have $D_t(f) = \Omega(t^{1/d})$ (resp. $\Omega((\log t)^{-\varepsilon})$).

Proof. Let $t \in \mathbb{N}$. If f is not preperiodic, then neither is its trace τ_f. By the Morse-Hedlund theorem, we must have $\mathcal{K}_{\tau_f}(t) > t$. By definition, D_t can be written as $2r + 1$ such that for all $x, y \in X$ such that $x_{[-r,r]} = y_{[-r,r]}$, we have $T_f(x)_{[0,t[} = T_f(y)_{[0,t[}$, so that $\mathcal{K}_{\tau_f}(t) \leq \mathcal{K}_X(D_t)$. We obtain $D_t \geq \mathcal{K}_X^{-1}(\mathcal{K}_{\tau_f}(t)) > \mathcal{K}_X^{-1}(t)$ because $\mathcal{K}_X$ is increasing.

Now, suppose that D_t is not $o(t)$, then subadditivity and the Fekete lemma imply that $D_t = \Theta(t)$ (and the same for $d(f^t)$). This argument is formalized for example in [6] or [1], and the limit of $D_t/2t$ corresponds to the maximal so-called Lyapunov exponent, in absolute value.   □

It is not known if there is the same lower gap for $d(f^t)$, or in general which kinds of growths are possible.

A natural object is the two-dimensional subshift of (two-sided) space-time diagrams of $f : \mathcal{X}_f = \{x \mid \forall t \in \mathbb{Z}, (x_{n,t})_{n \in \mathbb{Z}} = f((x_{n,t-1})_{n \in \mathbb{Z}}) \in X\}$. Following [7], we say that a vector line (or *direction*) $\ell \subset \mathbb{R}^2$ is expansive in $\mathcal{X}_f$ if there exists a width $r \in \mathbb{N}$ such that:

$$\forall x, y \in \mathcal{X}_f : (\forall v \in \mathbb{Z}^2 : d(v, \ell) < r \implies x_v = y_v) \implies x = y.$$

The following proposition is not difficult. It is for example a particular case of [1, Proposition 4.5].

Proposition 2. *A non-periodic automorphism f and its inverse are distorted if and only if $\mathcal{X}_f$ has the vertical direction as unique direction of nonexpansiveness.*

Actually, any 2D subshift is expansive in every nonvertical direction if and only if it is conjugate to $\mathcal{X}_f$ for some automorphism f such that both f and f^{-1} are periodic or distorted (a particular case of [1, Proposition 5.6]).

Note that Proposition 2 could motivate a notion of directional distortion, corresponding to endomorphisms whose space-times have a unique direction of nonexpansiveness, and whose composition with the corresponding shift is not prepe-riodic (in particular, if the unique direction of nonexpansiveness is irrational).

Several examples of such *extremely expansive* two-dimensional subshifts are known. A general self-simulating construction is given in [8], and effectivized (so that f is obtained as a partial local rule from the full shift) in [9,10].

We give in Sect. 4 a construction which is very similar to a second construction in [8], though independent. But first, in Sect. 3, we prove a link with one-head machines, which allows us to get distorted automorphisms of the full shift.

3   Distorted One-Head Machines

3.1   One-Head Machines

Let $\Delta = \{-1, +1\}$ be the set of directions. A *one-head machine* (or Turing machine) $\mathcal{M}$ is a triple (Q, Σ, δ) where Q is a finite set of states, Σ is a finite set of symbols, and $\delta \subset (Q \times \Delta \times Q) \sqcup (Q \times \Sigma \times Q \times \Sigma)$ is the transition function. This model (for example introduced in [11]) is equivalent to the one in [12], but handles reversibility better.

Noting $\tilde{\Sigma} = \Sigma \sqcup (Q \times \Sigma)$, where elements of $Q \times \Sigma$ are called *heads*, and $X_{\mathcal{M}} = \left\{ x\tilde{\Sigma}^{\mathbb{Z}} \mid \forall i, j \in \mathbb{Z}, x_i \in \Sigma \text{ or } x_j \in \Sigma \right\}$ (set of tapes with at most one head somewhere), we can associate to it the so-called *moving-head model* as the closed shift-invariant relation $M_{\mathcal{M}}$ of $X_{\mathcal{M}}$ defined by: $(x, x') \in M_{\mathcal{M}}$ if one of the following occurs:

$$x = x' \in \Sigma^{\mathbb{Z}};$$
$$\exists i \in \mathbb{Z}, (x_i, x'_i) \in \delta \text{ and } \forall j \neq i, x_j = x'_j \in \Sigma;$$
$$\exists i, i' \in \mathbb{Z}, (q, i' - i, q') \in \delta, x_i = (q, x'_i) \, ; \, x'_{i'} = (q', x_{i'}) \, ; \, \forall j \notin \{i, i'\}, x_j = x'_j \in \Sigma.$$

We actually focus on *total deterministic* machines, that is, machines where every configuration has exactly one successor, which makes $M_\mathcal{M}$ induce an endomorphism of $X_\mathcal{M}$, also noted $M_\mathcal{M}$. A *reversible one-head machine (RTM)* is a deterministic one-head machine for which $M_\mathcal{M}$ is actually an automorphism.

A (total deterministic) one-head machine $\mathcal{M}$ is *periodic* or *preperiodic* if the corresponding endomorphism is.

A configuration $x \in X_\mathcal{M}$ is *weakly periodic* if $M_\mathcal{M}^p(x) = \sigma^j(x)$ for some $p \geq 1$ and $j \in \mathbb{Z}$. We will say that it is *aperiodic* if it has no weakly periodic configuration containing a head[1] (that is, no configuration x with a nontrivial $p \geq 1$ and $q \in \mathbb{Z}$ such that $M_\mathcal{M}^p(x) = \sigma^q(x)$).

Fix a one-head machine $\mathcal{M} = (Q, \Sigma, \delta)$. If $x \in \tilde{\Sigma}^\mathbb{Z}$ contains a head, we write $s_t(x)$ for the number of distinct cells that the head of $\mathcal{M}$ visits in the first t steps starting from configuration x (taking the number of distinct cells rather than the position makes it nondecreasing, which simplifies some arguments). The function $m : \mathbb{N} \to \mathbb{N}$ defined by $m(t) = \max_x s_t(x)$ is called the *movement bound* of $\mathcal{M}$. The *speed* of $\mathcal{M}$ is defined in [14] as the limit of $m(t)/t$, which exists by subadditivity. A one-head machine will be called *distorted* if it is not periodic but m is sublinear in t.

Remark 2. It is easy to see that $D_t(M_\mathcal{M}) = m(t)$. In particular, $\mathcal{M}$ is distorted if and only if $M_\mathcal{M}$ is a distorted endomorphism.

We will prove that aperiodic one-head machines are examples of distorted machines.

3.2 Speed Trichotomy

In this section, we give some information on the possible speeds of one-head machines, namely that there are two gaps of impossible movements.

Theorem 1. *Let $\mathcal{M} = (Q, \Sigma, \delta)$ be a one-head machine with movement bound m. Then exactly one of the following holds:*

- *m is bounded ($\mathcal{M}$ is preperiodic);*
- *$m(t) = \Omega(\log t)$ and $m(t) = O(t/\log t)$ ($\mathcal{M}$ is distorted);*
- *$m(t) = \Theta(t)$ ($\mathcal{M}$ has positive speed).*

The preperiodic and positive speed cases are quite well understood. It can even be shown that some periodic configuration achieves the maximal speed [14]. We do not know what kinds of intermediate growth functions can be realized with distorted one-head machines.

Here is a simple counting lemma. If $H : \mathbb{N} \to \mathbb{N}$ is nondecreasing, we write $\lfloor H^{-1}(n) \rfloor$ for the largest ℓ such that $H(\ell) \leq n$.

Lemma 1. *Let $a_0, \dots, a_n$ be in $\mathbb{N}$ and suppose that $|\{i \in [\![0, n]\!] \mid a_i = \ell\}| \leq h(\ell)$ for all ℓ. We have the following.*

---

[1] This corresponds to classical aperiodicity in the so-called *moving-tape model* or *trace subshift* from [12,13].

1. *If* $H(\ell) = \sum_{i=0}^{\ell} h(i)$, *then* $\displaystyle\sum_{i=0}^{n} a_i \geq \sum_{i=0}^{\lfloor H^{-1}(n)\rfloor} ih(i)$.

2. *Moreover, if* $h(\ell) = \alpha^{\ell}$, *then* $\displaystyle\sum_{i=0}^{n} a_i = \Omega(n\log n)$.

Proof. 1. Define $b_i = k \iff i \in \llbracket H(k-1), H(k) \llbracket$ for $k \geq 0$. Then we have

$$\sum_{i=0}^{n} a_i \geq \sum_{i=0}^{n} b_i \geq \sum_{i=0}^{\lfloor H^{-1}(n)\rfloor} ih(i),$$

where the first inequality follows by sorting the a_i in increasing order and observing that then necessarily $a_i \geq b_i$ for all i, and the second follows by a direct counting argument.

2. If $\ell \leq \log_{\alpha} n - 1$, then $H(\ell) = \sum_{i=0}^{\ell} h(i) \leq \alpha^{\ell+1} \leq n$, so $\lfloor H^{-1}(n)\rfloor \geq \log_{\alpha} n - 2$. Thus

$$\sum_{i=0}^{n} a_i \geq \sum_{i=0}^{\lfloor H^{-1}(n)\rfloor} ih(i)$$
$$\geq \sum_{i=0}^{\log_{\alpha} n - 2} ih(i)$$
$$\geq (\log_{\alpha} n - 2)h(\log_{\alpha} n - 2)$$
$$= (\log_{\alpha} n - 2)\alpha^{\log_{\alpha} n - 2}$$
$$= (\log_{\alpha} n - 2)n\alpha^{-2}$$
$$= \Omega(n\log n).$$

$\square$

The upper bound is achieved by a counting argument, and the right object to count are the crossing sequences, which we now define.

To any machine $\mathcal{M}$, configuration $x \in X_{\mathcal{M}}$ and position $J \subset \mathbb{Z}$, we can associate the *crossing times* $\theta_J(x)$ as the ordered set of times $k \in \mathbb{N}$ such that $\exists i \in J, M_{\mathcal{M}}^k(x)_i \in Q \times \Sigma$; it is formally a tuple, but sometimes we use set notation, like its cardinality $|\theta_J(x)|$ or diameter $\max \theta_J(x) - \min \theta_J(x)$. Moreover for all steps $t \in \mathbb{N}$, we can associate the (partial) *crossing sequence* $u_{J,t}(x) = (M_{\mathcal{M}}^k(x)_J)_{k \in \theta_{J,t}(x)} \in (\tilde{\Sigma}^J)^*$, where $\theta_{J,t}(x) = \theta_i(x) \cap \llbracket 0, t \rrbracket$. This definition is close to a finitary version of the notion in [14,15], except we take the sequence at a given cell rather than between two neighboring cells, which makes no difference except for writing. We use notations θ_i, $\theta_{i,t}$, $u_{i,t}$ if $J = \{i\}$.

We are now ready to prove the main equivalence of this section.

Proposition 3. *Let* $\mathcal{M}$ *be a one-head machine. The following are equivalent.*

1. $m(n)$ *is not* $O(n/\log n)$.

2. There exist a configuration $x \in X_{\mathcal{M}}$, two distinct positions $i, j \in \mathbb{Z}$, and a step $t \in \mathbb{N}$ such that $u_{i,t}(x) = u_{j,t}(x)$ are nonempty.
3. There exists a configuration $x \in X_{\mathcal{M}}$ such that the cardinality of $\theta_i(x)$ is uniformly bounded for $i \in \mathbb{Z}$.
4. There exists a configuration $x \in X_{\mathcal{M}}$ such that the diameter of $\theta_i(x)$ is uniformly bounded for $i \in \mathbb{Z}$.
5. There exists a weakly periodic configuration which is not periodic.
6. $\mathcal{M}$ has positive speed: $m(n) = \Omega(n)$.

Point 2 actually remains equivalent if the first visited crossing sequence admits the other one as a prefix. The implications $2 \Rightarrow 3$ (resp. $3 \Rightarrow 4$) could also have been derived from looking at the countable-state Markov shift built in [14] (resp. from a general result over path spaces [16]), but we give specific proofs for completeness.

Proof. $1 \Rightarrow 2$ Let x be a configuration and t a step, and $J = \{i \in \mathbb{Z} | \theta_{i,t}(x) \neq \emptyset\}$ the set of visited cells. Since $[\![0, t[\![= \bigsqcup_{i \in J} \theta_{i,t}(x)$, we get that $t = \sum_{i \in J} |\theta_{i,t}|$, which is the sum of lengths of the crossing sequences. Suppose that every crossing sequence $u_{i,t}(x)$ is distinct, for $i \in J$. There are at most $(|\Sigma| |Q|)^{\ell}$ distinct crossing sequences of length ℓ, so it follows from Lemma 1 that $t = \Omega(|J| \log |J|)$. We get:

$$m(t) = \sup_x s_t(x) = \sup_x s_{\gamma(t) \log \gamma(t)}(x) \leq B\gamma(t) \leq 2B \frac{t}{\log t},$$

for some constant B and all large enough t, and where γ is the inverse of the function $t \mapsto t \log t$, which satisfies $\gamma(t) \leq \frac{t}{\log t - \log \log t} \leq 2\frac{t}{\log t}$ for large t.

$2 \Rightarrow 3$ By symmetry, we can assume that $j > i$ and $\min \theta_{j,t}(x) > \min \theta_{i,t}(x)$. By shifting and applying $M_{\mathcal{M}}$, we can assume that $i = 0$ and $\min \theta_{i,t}(x) = 0$. We can also assume that t is minimal for the property that $u_{0,t}(x) = u_{j,t}(x)$ is not empty. Equivalently, t is the first step n for which $k_n = |\theta_{0,n}(x)| - |\theta_{j,n}(x)| = 0$. Since $k_0 = 1$ and for $n \in \mathbb{N}$, $k_{n+1} \in k_n + \{-1, 0, 1\}$, we get that $k_n > 0$ if $n < t$, which means that $t' = \max \theta_{0,t}(x) < t = \max \theta_{j,t}(x)$. Note that $t + 1 \in \theta_{j+1}(x)$, because $M_{\mathcal{M}}^t(x)_j = M_{\mathcal{M}}^{t'}(x)_0$ gives a right movement by δ (if the machine head had been going to the left on 0 at $t' = \max \theta_{0,t}(x)$, then it could not have reached position j before time t).

Let $J_{-1} =]\!] - \infty, 0]\!]$ and, for $n \in \mathbb{N}$, $J_n = jn + [\![0, j]\!]$. Let $y \in X_{\mathcal{M}}$ have a tape that is periodic on the right in the following way: $y_{J_{-1}} = x_{J_{-1}}$ and $y_{jn +]\!]0,j]\!]} = x_{]\!]0,j]\!]}$.

Let us build inductively a nondecreasing map $\phi : \mathbb{N} \times (\mathbb{N} \sqcup \{-1\}) \to [\![0, t+1]\!]$ such that for all steps $k \in \mathbb{N}$ and $n \in \mathbb{N} \sqcup \{-1\}$, if $\theta_{J_n, k}(y)$ is nonempty then $M_{\mathcal{M}}^k(y)_{J_n} = M_{\mathcal{M}}^{\phi(k,n)}(x)_{J_{\min(n,0)}}$; besides, $\phi([\![0, k]\!] \times \{n\}) = \{0\} \cup \theta_{J_{\min(n,0)}, \phi(k,n)} \cup (\theta_{J_{\min(n,0)}, \phi(k,n)} + 1)$ (in particular, this gives $u_{J_n, k}(y) = u_{J_{\min(n,0)}, \phi(k,n)}(x)$); moreover, the restriction of ϕ to $\theta_{J_n, k}(y) \times \{n\}$ is an injection onto $\theta_{J_{\min(n,0)}, \phi(k,n)}(x)$.

$\phi(0, n) = 0$ clearly satisfies this. Now, suppose that ϕ has been built up to step $k \in \mathbb{N}$ for all $n \in \mathbb{N} \sqcup \{-1\}$. Let $n \in \mathbb{N} \sqcup \{-1\}$.

- If $k, k+1 \notin \theta_{J_n}(y)$, then $\phi(k+1, n) = \phi(k, n)$ is satisfying because J_n is unchanged at this moment.

- If $k \in \theta_{J_n}(y)$, then it is clear that $\phi(k+1, n) = \phi(k, n) + 1$ satisfies the two properties. By hypothesis, $M_{\mathcal{M}}^k(y)_{J_n} = M_{\mathcal{M}}^{\phi(k,n)}(x)_{J_{\min(n,0)}}$, so that $\phi(k, n) \in \theta_{J_{\min(n,0)}}(x)$. By the first remark of the proof, we see that $\phi(k, n) \neq t + 1$.

- Now if $k+1 \in \theta_{J_n}$ but $k \notin \theta_{J_n}$, then one can note that $k+1 \in \theta_{jn} \cup \theta_{j(n+1)}$ (the head should be in one boundary), say $k + 1 \in \theta_{jn}$ (the other case can be dealt with by symmetry), and in that case $k \in \theta_{jn-1}$, so that we have already defined $\phi(k+1, n-1) = \phi(k, n-1) + 1$, which is at most t (by hypothesis and because $\phi(k+1, n-1) \in \theta_{J_{n-1}}(y)$). We know that $M_{\mathcal{M}}^{k+1}(y)_{J_{n-1}} = M_{\mathcal{M}}^{\phi(k+1,n-1)}(x)_{J_{\min(n,0)}}$. In particular, $M_{\mathcal{M}}^{k+1}(y)_{jn} = M_{\mathcal{M}}^{\phi(k+1,n-1)}(x)_0$ contains the head, and the main hypothesis gives a corresponding time $\phi(k+1, n) \leq t$ such that $u_{j,\phi(k+1,n-1)}(x) = u_{0,\phi(k+1,n)}(x)$. From the construction of the previous steps, $\phi(k, n)$ is either 0 if $\theta_{J_n,k}$ is empty, or $\phi(\max \theta_{J_n,k}, n) + 1 = \max \theta_{J_n,\phi(k,n)}(x) + 1 < \phi(k+1, n)$ otherwise. In both cases, we get the wanted properties.

The last property of ϕ gives that for all $n \in \mathbb{N}$, $|\theta_{J_n,k}| = \left| \theta_{J_{\min(n,0)},\phi(k,n)} \right|$ and the fact that the map is bounded gives that this is at most $\left| \theta_{J_{\min(n,0)},t} \right| \leq t$. We obtain that the number of visited cells in the first k steps on y is $\Omega(jk/t)$.

$3 \Rightarrow 4$ Let $\ell \geq 1$ be minimal such that there is a configuration $x \in X_{\mathcal{M}} \backslash \Sigma^{\mathbb{Z}}$ such that $\forall i \in \mathbb{Z}, |\theta_i(x)| \leq \ell$. Assume that for all $n \in \mathbb{N}$, there exists $i_n \in \mathbb{Z}$ for which the diameter of $\theta_{i_n}(x)$ is at least n. Let us consider a limit point y of $(M_{\mathcal{M}}^{\min \theta_{i_n}(x)} \sigma^{i_n}(x))_{n \in \mathbb{N}}$. By minimality of ℓ, we know that $|\theta_i(y)| \geq \ell$ for some $i \in \mathbb{Z}$. Let $t = \max \theta_i(y)$, and n be such that for all $m \geq n$, $\forall k \leq t, M_{\mathcal{M}}^k \sigma^{im}(x)_i = M_{\mathcal{M}}^k(y)_i$. In particular, $u_{i,t}(\sigma^{im}(x)) = u_{i,t}(y)$ has length at least ℓ. By assumption, it actually has length ℓ and $\theta_i(\sigma^{im}(x)) \subset [\![0, t]\!]$. For every $m \geq \max(n, t)$, we have $\max \theta_{i_m}(x) \geq m \geq t$, so that, after time t, the head is in the connected component of $\mathbb{Z} \backslash \{i + i_m\}$ that contains i_m. Let us assume that $i > 0$ (the argument is symmetric), so that this connected component is $]\!]-\infty, i+i_m[\![$ for every m, and let j be one position taken by the head after time t. Then for all $m \geq \max(n, t)$, $j < i+i_m$, which means $i_m > j - i$. On the other hand, if i_m is itself a position that the head takes after time t, so it must be in $]\!]-\infty, i+i_{\max(n,t)}[\![$. It results that $\{i_m \mid m \geq \max(n, t)\}$ is included in the finite set $]\!]j-i, i+i_{\max(n,t)}[\![$, which contradicts its infinity.

$4 \Rightarrow 5$ Let $n \geq 1$ and $x \in X_{\mathcal{M}} \backslash \Sigma^{\mathbb{Z}}$ such that for all $i \in \mathbb{Z}$, the diameter of $\theta_i(x)$ is at most n. By the pigeonhole principle, there are two distinct positions $i, j \in \mathbb{Z}$ such that $\theta_i(x) - \min \theta_i(x) = \theta_j(x) - \min \theta_j(x)$ and $u_i(x) = u_j(x)$. Assume that $i = 0 < j$ and $\min \theta_i(x) = 0 < p = \theta_j(x)$ (by symmetry). Then our assumption says that $(M_{\mathcal{M}}^t(x)_i)_{t \in \mathbb{N}} = (M_{\mathcal{M}}^t(x)_j)_{t \in [\![p,\infty[\![}$. Let us define $y \in X_{\mathcal{M}}$ by $y_{jm+n} = M_{\mathcal{M}}^{(\max(0,-pm)}(x)_n$ if $m \in \mathbb{Z}$ and $n \in [\![1, j]\!]$. A standard cellular automata argument (an a little drawing) can convince that the pieces of space-time diagrams fit, so that, by induction on $t \in [\![0, p]\!]$: if

$n \in [\![1, j]\!]$ and $m > 0$, then $M_{\mathcal{M}}^t(y)_{jm+n} = x_n = y_{j(m-1)+n}$, and if $m \leq 0$, $M_{\mathcal{M}}^t(y)_{jm+n} = M_{\mathcal{M}}^{t-pm}(x)_n$. In particular, we get that $M_{\mathcal{M}}^p(y)_{jm+n} = y_{j(m-1)+n} = \sigma^j(y)_{jm+n}$.

$5 \Rightarrow 6$ It is clear that any configuration x such that $M_{\mathcal{M}}^p(x) = \sigma^j(x)$ has a speed $s_t(x) \sim jt/p$.

$6 \Rightarrow 1$ This is obvious.

$\square$

Proof (of Theorem 1). From Proposition 1 together with Remark 2, we know that if $\mathcal{M}$ is not preperiodic, then $m(t) = \Omega(\log t)$. The other gap corresponds to the implication $1 \Rightarrow 6$ in Proposition 3. $\square$

3.3 Aperiodic Machines

Theorem 2. *Every aperiodic one-head machine is distorted.*

In particular, there exist distorted one-head machines: see for example [3,17] for constructions of aperiodic machines. The latter is even minimal in the moving tape model (which directly implies aperiodicity, except over the trivial alphabet).

Proof. Consider the three cases of Theorem 1. If $\mathcal{M}$ is aperiodic, it naturally cannot be preperiodic. If $\mathcal{M}$ were in the last case of the theorem, its trace subshift would contain a periodic point $y \in (Q \times \Sigma)^{\mathbb{N}}$ with positive speed. On a configuration where this movement is realized, every cell is visited a bounded number of times, during a time interval of bounded length. Thus $\mathcal{M}$ is essentially performing a finite transduction, and it is easy to extract, by the pigeonhole principle, a configuration in $Q \times \Sigma^{\mathbb{Z}}$ where $\mathcal{M}$ acts periodically. $\square$

The machine constructed in [3] also has the property that the trace subshift of the one-head machine (the subshift encoding possible sequences of states that the head can enter when acting on a configuration) has a substitutive structure, and an explicit substitution is given. As the head movement only depends on the trace, it should be possible to compute the movement bound explicitly using spectral properties of the matrix associated to the substitution (see [18]), but this requires a bit of work since the substitution given in [3] is not primitive.

3.4 Distortion on Sofic Shifts

The question of distortion is most interesting on simple subshifts, as then distortion comes from the automorphism itself and not the structure of the subshift. In [1], it is stated in particular that it is not known whether range-distortion can be achieved on transitive subshifts. In this section, we show that the existence of distorted one-head machines directly implies the existence of distorted automorphisms on all uncountable sofic shifts.

The following lemma is a direct corollary of the construction in [19, Lemma 7] (the result is proved for mixing SFTs in [20], with essentially the same construction).

Lemma 2. *Let X be a full shift and Y an uncountable sofic shift. Then there exist $C, B \in \mathbb{N}$ and an embedding ϕ from the endomorphism monoid of X to that of Y such that $|D(\phi(f)) - BD(f)| \leq C$ for all endomorphisms f of X.*

The number B comes from the fact that individual (pairs of) letters are written as words of length B occurring in Y, and C comes from the fact that the rule is only applied in "safe contexts".

Lemma 3. *Let $\mathcal{M} = (Q, \Sigma, \delta)$ be a deterministic total one-head machine. Then, letting*

$$\Gamma = ((\Sigma^2 \times \Delta) \cup (Q \times \Sigma) \cup (\Sigma \times Q)),$$

there is a cellular automaton $f : \Gamma^{\mathbb{Z}} \to \Gamma^{\mathbb{Z}}$ such that if $m : \mathbb{N} \to \mathbb{N}$ is the movement bound for $\mathcal{M}$, then $D(f^t) \leq m(t)$ for all $t \in \mathbb{N}$. Moreover, f is reversible (resp. preperiodic) if and only if $\mathcal{M}$ is.

The proof uses so-called 'conveyor belts' to deal with configurations with several heads. One could also use the construction of [2] to embed the one-head machine to a cellular automaton, and obtain the same result.

Proof (of Lemma 3). The proof is similar to that of [19, Lemma 7]. For a residual set of points $x \in \Gamma^{\mathbb{Z}}$, we can split x into a product $x = \cdots w_{-2} w_{-1} w_0 w_1 w_2 \cdots$ such that for each $i \in \mathbb{Z}$, we have

$$w_i \in (\Sigma^2 \times \{+1\})^*((Q \times \Sigma) \cup (\Sigma \times Q))(\Sigma^2 \times \{-1\})^* \cup (\Sigma^2 \times \{+1\})^*(\Sigma^2 \times \{-1\})^*,$$

and this factorization is clearly unique: every point in $\Gamma^{\mathbb{Z}}$ can be seen as a point of this form, but the leftmost and/or rightmost words can be degenerate, and have an infinite number of ± 1. It is enough to define how f transforms these words, and if the resulting map is uniformly continuous (which will be evident from the construction), then f uniquely extends to a continuous function on the full shift. Shift-commutation follows automatically because the decomposition of x is unique and the decomposition process is shift-invariant, and thus we obtain a cellular automaton.

On words in $(\Sigma^2 \times \{+1\})^*(\Sigma^2 \times \{-1\})^*$, we do nothing. If $w \in (\Sigma^2 \times \{+1\})^*((Q \times \Sigma) \cup (\Sigma \times Q))(\Sigma^2 \times \{-1\})^*$, let $w' \in (\Sigma^2)^*((Q \times \Sigma) \cup (\Sigma \times Q))(\Sigma^2)^*$ be the word obtained from w by erasing the arrows. We see w' as a 'conveyor belt', wrapped around which is a word of length $2|w'|$. More precisely, let $u = \pi_1(w')$ and v equals the reversal $\overline{\pi_2(w')}$, of $\pi_2(w')$, and observe that one of these words is in Σ^+ and the other one is in $\Sigma^* Q \Sigma^*$.

Apply the transition function of the one-head machine to the configuration $(uv)^{\mathbb{Z}}$. Note that this configuration contains infinitely many heads, but as they move with the same rule, the movement is still well-defined. Note also that if the machine $\mathcal{M}$ is reversible, then this application is reversible as well, in the sense that the inverse of $\mathcal{M}^{-1}$ applied at every head undoes the transition step of $\mathcal{M}$ even on this periodic configuration. (This justifies the last sentence in the statement of the lemma.)

Now, the resulting configuration $(u'v')^{\mathbb{Z}}$ still contains exactly one head in every pattern $u'v'$. This configuration was obtained by a bijection that unwrapped a word $w' \in (\Sigma^2)^*((Q \times \Sigma) \cup (\Sigma \times Q))(\Sigma^2)^*$ to a pair of words. Perform the inverse of this bijection, rewrapping u' and v' to a word in $(\Sigma^2)^*((Q \times \Sigma) \cup (\Sigma \times Q))(\Sigma^2)^*$, and add a $+1$ and -1 component pointing towards the machine head to each cell containing a symbol in Σ^2. This defines f. Note that x and $f(x)$ always have the same decomposition, and if the one-head machine is reversible, its reverse one-head machine defines f^{-1}, so f is reversible if $\mathcal{M}$ is.

To see that $D(f^t) \leq \tilde{m}(t)$, consider any configuration $x \in \Gamma^{\mathbb{Z}}$. If there is no machine head in $x_{[-m(t),m(t)]}$, then $f^t(x)_0 = x_0$, since no machine head can travel by more than $m(t)$ cells in t steps. If there is a machine head in this interval in some coordinate $j \in \mathbb{Z}$, we start simulating its movement (also modifying the tape according to its movement). Note that the one-head machines stay neatly in their separate conveyor belts, so no machines crash into each other during this simulation. If a head steps out of the interval $[-m(t), m(t)]$ during the simulation, we can stop simulating it, as it will not reach the origin. After simulating heads for t steps, we know the value of $f^t(x)_0$. (Of course, we really only have to simulate a head if its conveyor belt contains the origin, and there is a unique such a head, but it does not hurt to simulate all of them.)   □

Theorem 3. *Let X be an uncountable sofic shift. Then there exists a distorted automorphism on X.*

Proof. Let $\mathcal{M}$ be a distorted one-head machine. Then the cellular automaton f constructed in the previous lemma is distorted. By Lemma 2, we obtain the same cellular automaton on any uncountable sofic shift.   □

3.5   Undecidability of Distortion

In this section, we show that distortion is undecidable.

Theorem 4. *It is undecidable, given a reversible one-head machine $\mathcal{M}$, whether $\mathcal{M}$ is distorted.*

We can actually see from the proof (and from the reduction in [2]) that it is Π_1^0-complete.

Proof. Every one-head machine lies in exactly one of the three cases of Theorem 1. We have a semialgorithm for the periodic case (by simply computing powers of $\mathcal{M}$ and checking whether they are the identity map), and we have a semialgorithm for the case when $\mathcal{M}$ has positive speed by the computability of speed, presented in [14, Theorem 2.7]. If we had a semialgorithm for detecting distortion, we would then be able to decide all three classes, contradicting the undecidability of periodicity, established in [2, Theorem 8].   □

Corollary 1. *For every uncountable sofic subshift X, it is undecidable, given an automorphism Φ of X, whether Φ is distorted.*

Proof. Lemmas 2 and 3 are effective: if some reversible one-head machine is given, an automorphism of X can be built such that the radius growth differs at most by a multiplicative constant. If we could decide whether the corresponding automorphism is distorted, we could then decide whether the original machine is, which contradicts Theorem 4.   □

4   Unbalanced Distortion in General Subshifts

We give a general construction of a distorted automorphism. The distortion we aim for is 'highly unbalanced'.

As mentioned above, it is desirable that $N(f)$ is generated by a single interval, that is, $I(f) \in N(f)$. Thus the size of the difference between $I(f)$ and the minimal intervals in $N(f)$ somehow measures the 'badness' of $N(f)$. We give an automorphism where this difference grows fast along iterations of f: The following theorem shows that we can have $d(f^t)$ be close to logarithmic, while $D(f^t)$ is close to linear.

We note that our construction is very similar to a construction in [8], though ours is not (at least consciously) based on it. Our proof is based on self-similar mud machinery that allows the construction of tracks that take a long time to walk over, but return to their original state once passed. To organize the behavior required in the theorem is then not difficult, though getting the numbers right requires some care because ϕ and ψ are arbitrary.

Theorem 5. *Let* $\phi : \mathbb{N} \to \mathbb{N}$ *be any sublinear function and let* $\psi : \mathbb{N} \to \mathbb{N}$ *be any nondecreasing superlogarithmic function. Then there exist a subshift X and an automorphism $f : X \to X$ such that there exist arbitrarily large t_i such that* $d(f^{t_i}) \leq \psi(t_i)$ *but* $D(f^{t_i}) \geq \phi(t_i)$.

By Proposition 1, the function ϕ cannot be made linear and ψ cannot be made logarithmic from a subshift X with subexponential complexity.

The following is shown in [1, Theorem 3.24]. Let us say that f is weakly periodic if there exist $p \geq 1$ an $j \in \mathbb{Z}$ such that $f^p = \sigma^j$.

Theorem 6. *Let X be an SFT and $f : X \to X$ an automorphism which is not weakly periodic. Then $I(f^t) \in N(f^t)$ for all large enough t.*

They ask [1, Question 3.26] whether the assumption that X is an SFT is needed. Theorem 5 answers by showing a general subshift and an automorphism of it for which at infinitely many t the interval $I(f^t)$ is arbitrarily close to logarithmic in size, but all contiguous neighborhoods are arbitrarily close to linear in size.

The proof of Theorem 5 first needs a technical but simple lemma. A function $\phi : \mathbb{N} \to \mathbb{R}$ is sublinear if $\phi(t) = o(t)$. Functions with this property can have weird local behavior, which complicates the argument. We show that all sublinear functions are majored by sublinear functions with some additional nice properties.

A nondecreasing function $\psi : \mathbb{N} \to \mathbb{R}$ has asymptotic slope zero if $|\psi(t+1) - \psi(t)|$ tends to zero as $t \longrightarrow \infty$. Note that if for a function ϕ, we write $\partial \phi : \mathbb{N} \to \mathbb{R}$

for its discrete derivative $\partial\phi(t) = \phi(t+1) - \phi(t)$, then asymptotic slope zero, for an increasing function means just that $\partial\phi(t)$ tends to 0 as $t \longrightarrow \infty$. If $\psi : \mathbb{R} \to \mathbb{R}$ is piecewise linear and it is linear on every interval $[\![i, i+1]\!]$ where $i \in \mathbb{N}$, then the restriction $\psi : \mathbb{N} \to \mathbb{R}$ has asymptotic slope zero if and only if the slopes of the linear pieces of ψ tend to zero.

Lemma 4. *If $\phi : \mathbb{N} \to \mathbb{R}$ is sublinear, then there is a sublinear increasing piecewise linear function $\psi : \mathbb{R} \to \mathbb{R}$ with asymptotic slope zero such that $\psi(t) \geq \phi(t)$ for all $t \in \mathbb{N}$.*

Proof. First, we may assume that ϕ is nondecreasing, by replacing $\phi(t)$ with $\max_{i=0}^t \phi(t)$, as the resulting function stays sublinear.

By sublinearity, for all $k \geq 1$ there exists t_k such that $t \geq t_k \Rightarrow \phi(t) \leq t/k$. Pick such $t_k \in \mathbb{N}$ for all $k \geq 1$, and observe that we can increase any of the t_k without changing their relevant property. Thus, we can assume the following further properties:

- t_k is increasing in k,
- $t_{k+1}/k > t_k/(k-1)$ for $k \geq 2$.

Consider the sequence of points $(t_{k+1}, t_{k+1}/k) \in \mathbb{N} \times \mathbb{R}$. The second item makes sure that this sequence of points increases on the first axis, and the second makes sure the sequence also increases on the second axis. On $[\![t_2, \infty[\![$, define ψ as the piecewise linear function obtained by linearly interpolating values in the interval $[\![t_k, t_{k+1}]\!]$ between $[\![t_k/(k-1), t_{k+1}/k]\!]$.

The function ψ is now increasing $[\![t_2, \infty[\![$, because the point $(t_{k+1}, t_{k+1}/k)$ is strictly below point $(t_{k+2}, t_{k+2}/(k+1))$ for all $k \geq 1$.

Note that in the interval $[\![t_k, \infty[\![$ where $k \geq 2$, we have $\psi(t) \leq t/(k-1)$: Each of the points $(t_\ell, t_\ell/(\ell-1))$ for $\ell > k$ are strictly below the line $L_k = \{(x, x/(k-1)) | x \in \mathbb{N}\}$ because $x/(k-1) > x/(\ell-1)$. Thus, interpolating linearly between these points, we obtain a path that stays under L_k, and values of $\psi(t)$ are by definition on this path. It follows from this that ψ is sublinear.

Next, observe that $\psi(t+1) \leq \psi(t) + \frac{1}{k}$ whenever $t \geq t_{k+1}$. This is because the slope of the line between $(t_{\ell+1}, t_{\ell+1}/\ell)$ and $(t_{\ell+2}, t_{\ell+2}/(\ell+1))$ is

$$\frac{t_{\ell+2}/(\ell+1) - t_{\ell+1}/\ell}{t_{\ell+2} - t_{\ell+1}} \leq \frac{t_{\ell+2}/\ell - t_{\ell+1}/\ell}{t_{\ell+2} - t_{\ell+1}} = 1/\ell,$$

for all $\ell \geq 1$. Thus, increasing t by one can increase the value of $\psi(t)$ by at most $1/k$ whenever $t \geq t_{k+1}$, since any such t fits in one of the intervals $[\![t_\ell + 1, t_{\ell+1}[\![$ where $\ell \geq k$.

Finally, we show that $\psi(t) \geq \phi(t)$ for all but finitely many t, from which the claim follows by choosing the first few values of ψ suitably, and then increasing other values by a constant. Suppose then that $k \geq 2$. On the interval $t \in [\![t_k, t_{k+1}[\![, \psi(t)$ is linearly interpolated between $(t_k, t_k/(k-1))$ and $(t_{k+1}, t_{k+1}/k)$. In particular the line between $(t_k, t_k/k)$ and $(t_{k+1}, t_{k+1}/k)$ is strictly below the graph of ψ. But $t \in [\![t_k, t_{k+1}[\![\Rightarrow \phi(t) \leq t/k$, implying that $(t, \psi(t))$ is below the the point $(t, t/k)$, thus below the graph of ψ. $\square$

With a slightly more careful construction, ψ could be made to have strictly nonincreasing first difference function (that is, such that the slopes of the linear pieces decrease from piece to piece), though we do not need this.

A function $\psi : \mathbb{N} \to \mathbb{R}$ is 2-nice if for all $C \in \mathbb{N}$, $\psi(t + C) \le 2\psi(t)$ for all but finitely many t.

Corollary 2. *If $\phi : \mathbb{N} \to \mathbb{N}$ is sublinear, then there is a sublinear nondecreasing 2-nice function $\psi : \mathbb{N} \to \mathbb{N}$ such that $\psi(t) \ge \phi(t)$ for all t.*

Proof. Seeing ϕ as a function $\phi : \mathbb{N} \to \mathbb{R}$, the previous lemma gives us a sublinear increasing $\psi : \mathbb{N} \to \mathbb{R}$ with asymptotic slope zero. It is easy to see (by separate easy proofs in the bounded and in the unbounded case) that any nondecreasing function $\psi : \mathbb{N} \to \mathbb{R}$ with asymptotic slope zero is 2-nice. If $\psi : \mathbb{N} \to \mathbb{R}$ is increasing and 2-nice, then $t \mapsto \lceil \psi(t) \rceil : \mathbb{N} \to \mathbb{N}$ is nondecreasing and 2-nice as well, and it clearly majors ψ, thus ϕ.   □

Proof (of Theorem 5). By the previous corollary, we may assume that ϕ is sublinear and 2-nice.

Take the alphabet $\{0, 1, 2, 3, 4, 5\} \times \{0, >, <\}$. The number $\{0, 1, 2, 3, 4, 5\}$ is the *mud state* and $<$ and $>$ are called *runners*. We construct a cellular automaton f that preserves the number of runners in every configuration with the property that on every f-invariant subshift, the map h flipping left and right runners is a time-symmetry for f, assuming that two runners never meet (that is, no word in $\{<<, <>, ><, >>\}$ appears in the configuration).

We only describe how the CA behaves when runners do not meet, as we will construct our subshift so that this does not happen. Our CA is composed of two CA, $f = g_2 \circ g_1$. The CA g_1 moves every occurrence of $<$ to the left and $>$ to the right. The CA g_2 maps

$$
\begin{array}{ll}
(0, >) \mapsto (2, >) & \qquad (1, <) \mapsto (5, <) \\
(1, >) \mapsto (3, <) & \qquad (0, <) \mapsto (4, >) \\
(2, <) \mapsto (0, >) & \qquad (5, >) \mapsto (1, <) \\
(3, >) \mapsto (1, >) & \qquad (4, <) \mapsto (0, <)
\end{array}
$$

and the local rule is filled arbitrarily so that this is a symbol permutation. It is a good idea to think of a left-going runner $<$ as already being on the left of the symbol, and $>$ as being on the right. We use a shorthand notation reflecting this[2], and write $(a, 0)$ as simply a, $(a, >)$ as $a>$ and $(a, <)$ as $<a$. We also write $>w$ and $w<$ when the mud state of the cell the runner is on is not important.

Now, the idea is the following: We call a word $w \in \{0, 1\}^*$ a *track* if, when a runner $>$ enters it from one side, it eventually goes out from the other side, leaving w in whatever state it was originally in, with the additional property that the number of times the mud state of a cell in w turns from zero 0 or 1 to

---

[2] We could just as well actually define our cellular automaton this way, but then tracks (defined below) move on the tape when a runner passes over them, which muddies up the global picture.

a symbol in $\{2,3,4,5\}$ is odd. The first property means that there exists t such that for all $a \in \{0,1,2,3,4,5\}$ we have

$$f^t(>0^{|w|}, aw) = (0^{|w|}>, aw)$$

and

$$f^t(0^{|w|}<, wa) = (<0^{|w|}, wa).$$

The importance of the second requirement will be clarified later. Note that on the first step, the runner is not yet on the support of the word, and on the last step, it is on the last symbol of the word w. To a track w, we associate its duration $t(w) \in \mathbb{N}$, which is the least t with the property above.

We show an example of a track. The word 01 is a track, and $t(01) = 4$, since (showing configurations from left to right) we have the evolutions

$$>01; \quad 2>1; \quad 2<3; \quad 0>3; \quad 01>$$

$$01<; \quad 0<5; \quad 4>5; \quad 4<1; \quad <01$$

where we see that the word returned to its original state, and both 0 and 1 changed their state an odd number of times. Clearly the composition of two tracks is a track, and it is easy to show by induction that if w is a track, then $0w1$ is a track as well, so we have a full Dyck language of tracks. One can check the formulas $t(uv) = t(u) + t(v)$ and $t(0w1) = 4 + 3t(w)$. For example

$$t(001011) = 4 + 3t(0101) = 4 + 3(t(01) + t(01)) = 28.$$

Just for fun, let us show how the head moves through 001011, representing configurations top-down, then left to right. We have

>001011	2012>11	2014>53	2<01013	0012>13
2>01011	2012<31	2014<13	0>01013	0012<33
22>1011	2010>31	201<013	02>1013	0010>33
22<3011	20101>1	20<5013	02<3013	00101>3
20>3011	20101<3	24>5013	00>3013	001011>
201>011	2010<53	24<1013	001>013	

and one can check that the vector recording the number of times each symbol 0 or 1 was changed is the all-odd vector $(1,3,3,3,3,1)$. The right-to-left case is symmetric.

Now fix $w_0 = 01$. Suppose w_i has been defined and define $w_{i+1} \in \{0,1\}^*$ as

$$w_{i+1} = (w_i 0)^{k_{i+1}} w_i (1w_i)^{k_{i+1}}$$

where $k_{i+1} \in \mathbb{N}$. Then, writing $\ell_j = |w_j|$ for all $j \in \mathbb{N}$, we have

$$\ell_{i+1} = (2k_{i+1} + 1)\ell_i + 2k_{i+1} \leq 4k_{i+1}\ell_i$$

if k_{i+1} is large enough as a function of ℓ_i (note that $\ell_i \geq 2$). Then w_{i+1} is a track, and writing $t_j = t(w_j)$ for all $j \in \mathbb{N}$, we have

$$t_{i+1} = 3^{k_{i+1}}(2t_i + 2) - t_i - 2 \geq 3^{k_{i+1}}t_i$$

if k_{i+1} is large enough. (The exact formula is provided for completeness, but we only need $t_{i+1} \geq 3^{k_{i+1}}t_i$ which obviously follows by induction from $t(0w1) = 4 + 3t(w)$.)

We pick k_{i+1} so that $\psi(\frac{1}{2}t_{i+1}) > 2\ell_{i+1}$. This is possible because when k_{i+1} grows, ℓ_{i+1} grows at a linear rate, and t_{i+1} grows exponentially, while ψ is superlogarithmic. More precisely, since $t_i > 2$ we have $\log_3 \frac{1}{2}t_{i+1} \geq k_{i+1}$ and $\ell_{i+1} \leq 4k_{i+1}\ell_i$. By the assumption on ψ, if n is large enough, we have $\psi(n) > 8\ell_i \log_3 n$, so in particular if k_{i+1} is large enough, we have

$$\psi(\frac{1}{2}t_{i+1}) > 8\ell_i \log_3 \frac{1}{2}t_{i+1} > 8\ell_i k_{i+1} \geq 2\ell_{i+1}.$$

We have obtained that t_i grows very fast as a function of ℓ, as it must make the function ψ – which can be arbitrarily close to logarithmic – overtake ℓ. Reversing our point of view, we have achieved that ℓ_i grows 'arbitrarily close to logarithmically' in t_i.

It is easy to prove by induction that for every $j < i$, we have a decomposition

$$w_i = w_j b_1 w_j b_2 \cdots b_m w_j$$

for some m, where the b_k are individual bits $b_k \in \{0, 1\}$.

Now, we construct our subshift X, which we call the *mud run subshift*. For each $i \in \mathbb{N}$, pick $q_i \in \mathbb{N}$ and $k_i \in \mathbb{N}$ so that $\phi(q_i t_i) < q_i \ell_i$ (which is true for any large enough q_i since ϕ is sublinear), and additionally so that $\frac{1}{2}t_{k_i} < \phi(q_i t_i) \leq t_{k_i}$, using the fact that $\phi(n + t_i) \leq 2\phi(n)$ for all large enough n. (Note that, if we pick $k_i \geq i + 1$, the value of t_{k_i} is not determined by the values $k_1, \dots, k_i$, but rather values up to k_{k_i}, so it is easy to make sure that t_{k_i} is much larger than $\phi(q_i t_i)$, and we can then increase q_i to get $\phi(q_i t_i)$ in the desired interval.)

For each i, take the periodic points $x_i = (>w_i^{q_i})^{\mathbb{Z}}$. Then x_i is a temporally periodic point for f, and the length of its f-orbit is $q_i t_i$, while the length of its σ-orbit is $q_i \ell_i$. Let $\chi : \{0, 1, 2, 3, 4, 5\} \times \{0, <, >\} \to \{0, 1, 2, 3, 4, 5\} \times \{0\}$ be the map that removes runners, and define the subshift Y as the closure of

$$\{\sigma^a(f^b(x_i)), \sigma^a(f^b(\chi(x_i))) \mid a \in \mathbb{Z}, b \in \mathbb{Z}, i \in \mathbb{N}\}.$$

It is easy to see that f is still an automorphism of this limit subshift (since it has the same inverse), and that every point in Y that is not in the $\{\sigma, f\}$-orbits of x_i has at most one runner (simply because $\ell_i \to \infty$).

Finally, define $X = Y \times \{0, 1\}^{\mathbb{Z}}$ and modify the behavior of f so that it behaves as before on Y, but additionally flips the bit on the second track whenever it turns a symbol from 0 or 1 to another symbol on the Y-component. Then by the assumption that we originally made for tracks that 0 and 1 are changed to another symbol an odd number of times, we have that for all $i \in \mathbb{N}$ and $z \in \{0,1\}^{\mathbb{Z}}$ we have $f^{q_i t_i}(x_i, z) = (x_i, z')$ where $z'_j = 1 - z_j$ for all $j \in \mathbb{Z}$.

We will now prove that the evolution of the neighborhoods of f on X has the properties we claim. More precisely, we pick a suitable sequence of times, $n_i = q_i t_i \in \mathbb{N}$, at which we look at the neighborhoods $N(f^{n_i})$. We show that due to our choice of the q_i, every interval in $N(f^{n_i})$ is of size at least $\phi(n_i)$. The reason for this is that already on the periodic points x_i generating X, we need neighborhoods of this size, since the runners move at a linear speed for a long time. We then show that simply due to the way the words w_i were constructed, we necessarily have

$$[-\psi(n_i), \infty[\, , \,] - \infty, \psi(n_i)]] \in N(f^{n_i}),$$

for these n_i. This is because, knowing the infinite tails, if we see no runners in those tails, we must be in a limit point, and in such points there is at most one runner, running at speed asymptotically slower than ψ. We do not have to look far to find it.

More precisely, for the first claim suppose $[a, b[\, \in N(f^{n_i})$. Then if $b - a + 1 < q_i \ell_i$, we see runners in $[a, b[$ in neither of the points $\sigma^{-a+1}(x_i)$ and $\sigma^{-a+1}(\chi(x_i))$. However, $f^{n_i}(\sigma^{-a+1}(x_i))_0 \neq \sigma^{-a+1}(x_i)_0$ and $f^{n_i}(\sigma^{-a+1}(\chi(x_i)))_0 \neq \sigma^{-a+1}(\chi (x_i))_0$. This contradicts $[a, b[\, \in N(f^{n_i})$. Since $\phi(n_i) = \phi(q_i t_i) < q_i \ell_i$, in particular every interval in $N(f^{n_i})$ is of length at least $\phi(n_i)$.

To see that $[-\psi(n_i), \infty[\, \in N(f^{n_i})$, note that if there are at least two runners in the $[-\psi(n_i), \infty[$-tail of a point $x \in X$, then runners appear periodically in x, and in fact we know precisely what will be at the origin after n_i steps. It remains to consider the limit set. It is enough to show that if x is in the limit set, and the unique runner on x is in cell 0, then after n_i steps it has moved by at most $\psi(n_i)$ steps (in either direction). Namely, if this is the case, then knowing the contents of $[-\psi(n_i), \psi(n_i)]]$ allows us to determine the contents of the cell 0 after the application of f^{n_i} by simply simulating the movement of the runner as long as it does not exit $[-\psi(n_i), \psi(n_i)]]$ (and if it does, the contents of cell 0 will no longer change).

To see this, consider the movement of a runner on one of the periodic points x_k. We claim that if $j < k$, then if a runner is in cell a of $f^n(x_k)$, then in $f^{n+t_j}(x_k)$, it will be in some cell in the interval $[a - 2\ell_j, a + 2\ell_j]$. To see this, recall that x_k is a concatenation of the words w_j separated by individual bits. Since tracks return to their original state after the runner has passed through them, in $f^n(x_k)$ a full intact copy of the track w_j appears in both of the intervals $[a, a + 2\ell_j]$ and $[a - 2\ell_j, a]$. It follows that exiting the interval $[a - 2\ell_j, a + 2\ell_j]$ requires passing through at least one complete track w_j, which takes time t_j.

Since f is continuous, the same will be true in the limit: every runner in a limit point x moves at most $2\ell_j$ cells in t_j time steps. Remember that we have $\frac{1}{2}t_{k_i} < n_i = q_i t_i \leq t_{k_i}$. In t_{k_i} time steps, the head moves at most $2\ell_{k_i}$ steps, so since $n_i = q_i t_i \leq t_{k_i}$, in n_i time steps we move a distance of at most

$$2\ell_{k_i} < \psi\left(\frac{t_{i+1}}{2}\right) \leq \psi(n_i)$$

cells, as required, where the last inequality follows because ψ is nondecreasing. This concludes the proof. □

5   Future Work

As mentioned in the section on one-head machines, we do not know much about the actual speeds of distorted machines. We tend to believe that the movement of the known examples is closer to $\log t$ than $t/\log t$, but have no rigorous proofs for any examples. It should be routine to compute the movement bound for the SMART machine, given that its trace subshift is known, and this would already clarify the situation quite a bit.

The group generated by reversible one-head machines is of prime interest and is the purpose of [21], though defined in a more natural way, which emphasizes its relation to the topological full group. In group theory, an element $g \in G$ is called *distorted* if powers of g grow sublinearly in the word norm of some finitely generated subgroup. We do not give examples of cellular automata or one-head machines that are distorted in this sense, but we do believe constructing them is possible.

Conjecture 1. The group generated by reversible one-head machines contains a distortion element.

More concretely, we find it plausible that the SMART machine is distorted. In the journal version of [21], which is in preparation, it will be shown that the subgroup of elementary one-head machines is finitely generated. Our conjecture is based on the fact that it seems plausible that a word expressing SMART in terms of these generators will not be linear, but doing this construction explicitly would presumably be a lot of work. If the conjecture is true, it follows (see Sect. 3.4) that automorphism groups of full shifts also have distortion elements.

References

1. Cyr, V., Franks, J., Kra, B.: The spacetime of a shift automorphism. ArXiv e-prints, October 2016
2. Kari, J., Ollinger, N.: Periodicity and immortality in reversible computing. In: Ochmański, E., Tyszkiewicz, J. (eds.) MFCS 2008. LNCS, vol. 5162, pp. 419–430. Springer, Heidelberg (2008). doi:10.1007/978-3-540-85238-4_34
3. Cassaigne, J., Ollinger, N., Torres-Avilés, R.: A small minimal aperiodic reversible turing machine. J. Comput. Syst. Sci. **84**, 288–301 (2017)
4. Hedlund, G.A.: Endomorphisms and automorphisms of the shift dynamical system. Math. Syst. Theory **3**, 320–375 (1969)
5. Cyr, V., Franks, J., Kra, B., Petite, S.: Distortion and the automorphism group of a shift. ArXiv e-prints, November 2016
6. Tisseur, P.: Cellular automata and lyapunov exponents. Nonlinearity **13**(5), 1547–1560 (2000)
7. Boyle, M., Lind, D.: Expansive subdynamics. Trans. AMS **349**(1), 55–102 (1997)

8. Hochman, M.: Non-expansive directions for $\mathbb{Z}^2$ actions. Ergodic Theory Dyn. Syst. **31**, 91–112 (2011)
9. Zinoviadis, C.: Hierarchy and Expansiveness in 2D Subshifts of Finite Type. PhD thesis, Turun Yliopisto, Turku (2015)
10. Guillon, P., Zinoviadis, C.: Hierarchy and expansiveness in two-dimensional subshifts of finite type, March 2016. draft
11. Morita, K.: Universality of reversible two-counter machine. Theoret. Comput. Sci. **168**(2), 303–320 (1996)
12. Kůrka, P.: On topological dynamics of turing machines. Theoret. Comput. Sci. **174**(1), 203–216 (1997)
13. Gajardo, A., Guillon, P.: Zigzags in turing machines. In: Ablayev, F., Mayr, E.W. (eds.) CSR 2010. LNCS, vol. 6072, pp. 109–119. Springer, Heidelberg (2010). doi:10. 1007/978-3-642-13182-0_11
14. Jeandel, E.: Computability of the entropy of one-tape Turing machines. ArXiv e-prints, February 2013
15. Hennie, F.: One-tape, off-line turing machine computations. Inf. Control **8**, 553–578 (1965)
16. Salo, V.: Subshifts with sparse projective subdynamics. ArXiv e-prints, May 2016
17. Blondel, V.D., Cassaigne, J., Nichitiu, C.: On the presence of periodic configurations in turing machines and in counter machines. Theoret. Comput. Sci. **289**, 573–590 (2002)
18. Adamczewski, B.: Balances for fixed points of primitive substitutions. Theoret. Comput. Sci. **307**(1), 47–75 (2003)
19. Salo, V.: A note on subgroups of automorphism groups of full shifts. Ergodic Theory and Dynamical Systems, pp. 1–13 (2016)
20. Kim, K.H., Roush, F.W.: On the automorphism groups of subshifts. Pure Math. Appl. **1**(4), 203–230 (1990)
21. Barbieri, S., Kari, J., Salo, V.: The group of reversible turing machines. In: Cook, M., Neary, T. (eds.) AUTOMATA 2016. LNCS, vol. 9664, pp. 49–62. Springer, Cham (2016). doi:10.1007/978-3-319-39300-1_5

Fast One-Way Cellular Automata
with Reversible Mealy Cells

Martin Kutrib, Andreas Malcher[✉], and Matthias Wendlandt

Institut für Informatik, Universität Giessen,
Arndtstr. 2, 35392 Giessen, Germany
{kutrib,malcher,matthias.wendlandt}@informatik.uni-giessen.de

Abstract. We investigate cellular automata that are composed of reversible components with regard to the recognition of formal languages. In particular, real-time one-way cellular automata (OCA) are considered which are composed of reversible Mealy automata. Moreover, we differentiate between three notions of reversibility in the Mealy automata, namely, between weak and strong reversibility as well as reversible partitioned OCA which have been introduced by Morita in [14]. Here, it turns out that every real-time OCA can be transformed into an equivalent real-time OCA with weakly reversible automata in its cells, whereas the remaining two notions seem to be weaker. However, a non-semilinear language is provided that can be accepted by a real-time OCA with strongly reversible cells. On the other hand, we present a context-free, non-regular language that is accepted by some real-time reversible partitioned OCA.

1 Introduction

The study of computational devices performing reversible computations is mostly motivated by the physical observation that a loss of information yields heat dissipation [12]. To avoid such situations computations are of interest in which every configuration has a unique successor configuration as well as a unique predecessor configuration so that at every point of the computation no information gets lost. Reversibility has been studied for many computational devices starting with Bennett's investigations for Turing machines in [3] where it is shown that for every (possibly irreversible) Turing machine an equivalent reversible Turing machine can be constructed. A similar result has been obtained for deterministic space-bounded Turing machines, in particular, deterministic linear bounded automata by Lange, McKenzie, and Tapp in [13]. For deterministic pushdown automata and deterministic queue automata the situation is different: in both cases it is possible to show (see, for example, [8,11]) that the reversible variant is weaker than the general model, that is, there are languages which can be accepted by the general model, but not by its reversible variant. In these cases, the loss of information in computations is inevitable. For deterministic multi-head finite automata the picture is split: for two-way multi-head finite automata Morita

© IFIP International Federation for Information Processing 2017
Published by Springer International Publishing AG 2017. All Rights Reserved
A. Dennunzio et al. (Eds.): AUTOMATA 2017, LNCS 10248, pp. 139–150, 2017.
DOI: 10.1007/978-3-319-58631-1_11

has shown in [17] that the general model and the reversible model coincide. On the other hand, in case of one-way motion it is shown in [9] that the reversible model is weaker than the general model. Reversible computations in deterministic finite automata have been introduced in [2] and it is shown in [20] that there are regular languages which cannot be accepted by any (one-way) reversible deterministic finite automaton. However, if two-way motion of the input head is allowed, it is known due to [5] that the general model and the reversible model coincide. A structural approach to reversible computing not depending on specific computational models has been proposed in [1].

For cellular automata (CA), the notion of reversibility has been investigated from several points of view. One fundamental result is that the reversibility of a cellular automaton is equivalent to the injectivity of the global transition function. Moreover, the injectivity of the global transition function is decidable for one-dimensional CAs, but becomes undecidable in higher dimensions. Details and literature for these results may be found in the survey paper [4]. The question whether every cellular automaton can be made reversible has been answered in the affirmative first by Toffoli who shows in [21] that every k-dimensional CA can be simulated by a $(k+1)$-dimensional reversible CA. This result has been improved by Morita and Harao in [19] where it is shown that every reversible Turing machine can be simulated by a one-dimensional reversible CA. By introducing the notion of partitioned cellular automata further improvements are given by Morita in [14,15] where, for example, the latter result is shown to hold also for one-dimensional *one-way* reversible CAs. More results on reversible CAs may be found in the survey paper [16].

In the context of language recognition, cellular automata are working on finite configurations with fixed boundary conditions. With regard to reversible computations, it is clear that reversibility in such devices cannot be defined on the injectivity of the global transition function. Thus, one considers computations that are reversible on the core of computation, namely, starting in the initial configuration and ending in the configuration given by the time complexity. From this point of view, language recognition by reversible devices has been studied for real-time two-way CAs [6], for real-time iterative arrays [7], and more recently for real-time one-way CAs [10]. Another recent result is provided by Morita in [18] where it is shown that every deterministic linear bounded automaton can be simulated by a reversible CA working on finite configurations with fixed boundary conditions.

In this paper, we consider another aspect of reversibility. In all cellular models studied so far the reversibility concerns configurations, that is, from every configuration the successor as well as the predecessor configuration can be computed in a unique way. Since CAs are basically arrays of interacting deterministic finite automata, one can also consider the reversibility of the single deterministic finite automata, that is, of the local transition function, and we will speak in this context of *locally reversible* CAs. For partitioned cellular automata and unbounded computations it is known [15,19] that such automata are globally reversible if and only if they are locally reversible if and only if they are locally injective. Apart from a theoretical interest in locally reversible computations, there is also a practical interest to investigate such CAs, since in this case the devices are composed of reversible components. Here, we will study local reversibility with regard to language recognition for weak cellular devices, namely, we will focus on real-time computations in one-way CAs (OCAs). Moreover, we consider OCAs having Mealy automata in their cells instead of deterministic finite automata. This generalization allows in particular a comparison with the notion of partitioned cellular automata. The paper is organized as follows. In Sect. 2 we summarize basic notions and introduce weakly and strongly reversible Mealy automata which are subsequently used to define one-way Mealy cellular automata with weakly or strongly reversible cells. Moreover, we provide an example of a non-semilinear language that is accepted by some real-time OCA with strongly reversible cells. Section 3 is devoted to investigating real-time OCAs with weakly reversible cells and it turns out that every real-time OCA can be converted to the former model. This means that every real-time OCA computation can be simulated by a real-time OCA with weakly reversible cells. In Sect. 4 we study reversible one-way partitioned CAs working in contrast to [14] on finite configurations with fixed boundary conditions. We first discuss how this notion is related to our concept of Mealy cellular automata. Then, it is shown that every regular language (reversible or not) is accepted by such automata. Moreover, it is possible to accept a certain context-free, non-regular language. Finally, we give a short conclusion. We would like to note that some proofs are omitted due to space considerations.

2   Preliminaries and Definitions

We denote the set of non-negative integers by $\mathbb{N}$. The reversal of a word w is denoted by w^R. For the length of w we write $|w|$. We write $\subseteq$ for set inclusion, and $\subset$ for strict set inclusion. In order to avoid technical overloading in writing, two languages L and L' are considered to be equal, if they differ at most by the empty word. Throughout the article two devices are said to be *equivalent* if and only if they accept the same language.

A deterministic finite Mealy automaton (DFMA) is a deterministic finite automaton that emits a symbol during each transition performed. So, it is particularly composed of a finite state set S, a finite input alphabet A, a finite output alphabet B, and a partial transition function δ that maps from $S \times A$ to $S \times B$. In this way, the new state and the symbol emitted during a transition are given. Let π_S denote the projection on the first component and π_B denote the projection on the second component of pairs from $S \times B$. A state in a DFMA is called a *sink state*, if the state can never be left once entered.

A DFMA is said to be *weakly reversible* (WREV-DFMA) if every pair (a, b) from $A \times B$ induces an *injective partial mapping* from the state set S to itself *via* the mapping $\delta_{(a,b)}: S \to S$ where $\delta_{(a,b)}(s) = s'$ if and only if $b = \pi_B(\delta(s, a))$ and $s' = \pi_S(\delta(s, a))$. In this case, the reverse transition function $\delta^\leftarrow: S \times A \times B \to S$ defined by $\delta^\leftarrow(s', a, b) = s$ if and only if $\delta(s, a, b) = s'$ induces for every pair

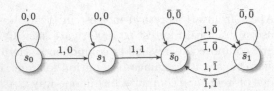

Fig. 1. A weakly reversible deterministic finite Mealy automaton, where an edge from p to q labeled by a, b means $\delta(p, a) = (q, b)$. Example transitions are $\delta(s_1, 1) = (\bar{s}_0, 1)$, $\delta_{(1,1)}(s_1) = \bar{s}_0$, $\delta^{\leftarrow}(\bar{s}_0, 1, 1) = s_1$ and $\delta^{\leftarrow}_{(1,1)}(\bar{s}_0) = s_1$.

(a, b) from $A \times B$ a (partial) injective function $\delta^{\leftarrow}_{(a,b)} : S \to S$ (see Fig. 1). A WREV-DFMA can also be considered as a (partial) permutation automaton.

A DFMA is said to be *strongly reversible* (SREV-DFMA) if every letter a from A induces an *injective partial mapping* from the state set S to itself *via* the mapping $\delta_a : S \to S$ where $\delta_a(s) = s'$ if and only if $s' = \pi_S(\delta(s, a))$. In this case, the reverse transition function $\delta^{\leftarrow}$ induces for every letter a from A a (partial) injective function $\delta^{\leftarrow}_a : S \to S$. The property of being strongly reversible is also known as being codeterministic.

Next, we consider one-way cellular automata whose cells are DFMAs. A one-way cellular automaton with Mealy cells (one-way Mealy cellular automaton) is a linear array of identical deterministic finite Mealy machines, called cells. Except for the rightmost cell each one is connected to its nearest neighbor to the right. The state transition of a cell depends on its current state and the latest output that has been emitted by its neighbor. We say that this output is the message sent to the neighbor. Initially, a distinguished initial message is sent. The rightmost cell receives information associated with a boundary symbol on its free input line. The state changes take place simultaneously at discrete time steps. The input mode for cellular automata is called parallel. One can suppose that all cells fetch their input symbol during a pre-initial step.

Formally, a *one-way Mealy cellular automaton* (OMCA) is a system given as $\langle S, F, A, B, \perp, \#, \delta \rangle$, where S is the finite, nonempty set of *cell states*, $F \subseteq S$ is the set of *accepting states*, $A \subseteq S$ is the nonempty set of *input symbols*, B is the finite, nonempty set of *messages*, $\perp \in B$ is the *initial message*, $\# \in B$ is the *boundary message*, and $\delta : S \times B \to S \times B$ is the *local transition function*.

A *configuration* of a one-way Mealy cellular automaton $\langle S, F, A, B, \perp, \#, \delta \rangle$ is a mapping $c : \{1, 2, \ldots, n\} \to (S \times B)$, for $n \geq 1$, that assigns a state and a message to each cell, where it is understood that the state is the current state of the cell and the message is the latest message sent by its neighbor. As before, given some $c(i) = (s, m)$, the projection on its state part s is denoted by $\pi_S(c(i))$ and the projection on its message part m is denoted by $\pi_B(c(i))$. The operation starts in a so-called *initial configuration*, which is defined by the given input $w = a_1 a_2 \cdots a_n \in A^+$. We set $c_0(i) = (a_i, \perp)$, for $1 \leq i \leq n - 1$, and $c_0(n) = (a_n, \#)$. Successor configurations are computed according to the global transition function Δ. Let c be a configuration with $n \geq 1$, then the successor configuration c' is

$$c' = \Delta(c) \iff \begin{cases} \pi_S(c'(i)) = \pi_S(\delta(c(i))), i \in \{1, 2, \ldots, n\} \\ \pi_B(c'(i)) = \pi_B(\delta(c(i+1))), i \in \{1, 2, \ldots, n-1\} \\ \pi_B(c'(n)) = \# \end{cases}.$$

An input w is accepted by a one-way Mealy cellular automaton M if at some time step during the course of its computation the leftmost cell enters an accepting state. The *language accepted by* M is denoted by $L(M)$. If all $w \in L(M)$ are accepted with at most $|w|+1$ time steps, then M is said to operate in *real-time*. The family of languages accepted by some device X operating in real-time is denoted by $\mathscr{L}_{rt}(X)$.

Note that the state transitions of cells in an OMCA depend on the current state and the latest output symbol emitted by the neighbor. So, taking a single cell as DFMA, its input alphabet is equal to its output alphabet B.

Now the structural restriction of one-way Mealy cellular automata we are interested in is that the single cells have to be *reversible* deterministic finite Mealy automata. These cellular automata are referred to by *one-way Mealy cellular automata with strongly or weakly reversible cells* and are denoted by SRC-OMCA and WRC-OMCA. In general, the transition functions for reversible deterministic finite Mealy automata may be partial in order to cope with situations that would drive the automaton into a rejecting sink state. Instead of entering the sink state, now the DFMA simply stops and rejects since it could not process the input entirely. However, since the concept of cellular automata does not allow single cells to stop, here, a rejecting sink state of the cells cannot be avoided in general. So, we slightly soften the notion of reversibility by disregarding rejecting sink states and say that an OMCA is an RC-OMCA if its cells are deterministic finite automata that are reversible with the exception of a possible rejecting sink state. However, it turns out in the next section that the disregarding of rejecting sink states is no restriction at least for WRC-OMCAs.

These definitions are justified and compared with related concepts after the following example that should clarify the notation.

Example 1. The non-semilinear language $\{a^n b^{k \cdot 2^n} \mid k, n \geq 1\}$ is accepted by the following one-way Mealy cellular automaton with strongly reversible cells $M = \langle S, F, \{a, b\}, B, \perp, \#, \delta \rangle$ in real time. We set $S = \{a, a_1, b, s_-, s_+\}$, where $B = \{1, 0, s, s_-, s_+, \perp, \#\}$, $F = \{s_+\}$, is the sole accepting state, and s_- is a rejecting sink state. The transition function δ is defined through:

1. $\delta(b, \perp) = (b, 1)$
2. $\delta(b, 1) = (b, 1)$
3. $\delta(b, \#) = (b, s)$
4. $\delta(b, s) = (b, s)$

5. $\delta(a, \perp) = (a, 0)$
6. $\delta(a, 0) = (a, 0)$
7. $\delta(a, 1) = (a_1, 0)$
8. $\delta(a, s) = (s_-, s_-)$
9. $\delta(a, s_-) = (s_-, s_-)$
10. $\delta(a, s_+) = (s_-, s_-)$

11. $\delta(a_1, 0) = (a_1, 0)$
12. $\delta(a_1, 1) = (a, 1)$
13. $\delta(a_1, s) = (s_+, s_+)$
14. $\delta(a_1, s_-) = (s_-, s_-)$
15. $\delta(a_1, s_+) = (s_+, s_+)$

and $\delta(s_-, x) = (s_-, s_-)$ and $\delta(s_+, x) = (s_-, s_-)$, for all $x \in B$.

By inspecting the transition function, it is verified that the cells are in fact SREV-DFMAs. An example computation on input a^3b^8 is depicted in Fig. 2. Basically, the consecutive b-cells stay in their state, where message 1 is transmitted in every step. Additionally, a message s is sent by the rightmost cell upon receiving the border message. This message moves through the b-cells one cell per time step.

The consecutive a-cells set up a binary counter with the least significant bit in the rightmost a-cell. To this end, state a is used to represent digit zero and state a_1 is used to represent digit one. A message 1 indicates a carry-over and a message 0 indicates no carry-over. Finally, when the signal s meets the a-cells, it becomes signal s_+ as long as it only sees digits one, that is, states a_1. Otherwise, it turns to signal s_- which is rejecting.

So, in order to accept an input, the leftmost cell has to enter state s_+. This is only possible if message s_+ has moved through a counter that represents a binary number of the form 1^n, that is, $2^n - 1$. Since due to the initial step, the counter starts to increase at time step one, this is only possible if message s has passed through a sequence of b-cells whose length is a multiple of 2^n. ∎

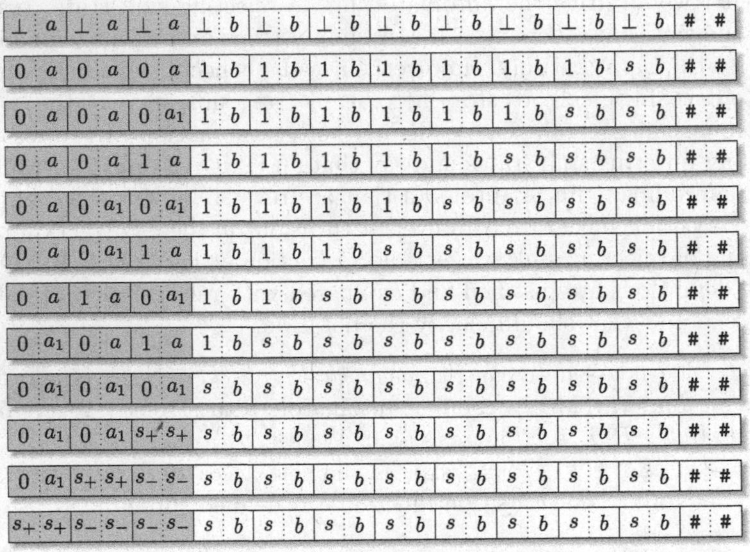

⊥ a	⊥ a	⊥ a	⊥ b	⊥ b	⊥ b	⊥ b	⊥ b	⊥ b	⊥ b	⊥ b	#	#
0 a	0 a	0 a	1 b	1 b	1 b	1 b	1 b	1 b	1 b	s b	#	#
0 a	0 a	0 a_1	1 b	1 b	1 b	1 b	1 b	1 b	s b	s b	#	#
0 a	0 a	1 a	1 b	1 b	1 b	1 b	1 b	s b	s b	s b	#	#
0 a	0 a_1	0 a_1	1 b	1 b	1 b	1 b	s b	s b	s b	s b	#	#
0 a	0 a_1	1 a	1 b	1 b	1 b	s b	s b	s b	s b	s b	#	#
0 a	1 a	0 a_1	1 b	1 b	s b	s b	s b	s b	s b	s b	#	#
0 a_1	0 a	1 a	1 b	s b	s b	s b	s b	s b	s b	s b	#	#
0 a_1	0 a_1	0 a_1	s b	s b	s b	s b	s b	s b	s b	s b	#	#
0 a_1	0 a_1	s_+ s_+	s b	s b	s b	s b	s b	s b	s b	s b	#	#
0 a_1	s_+ s_+	s_- s_-	s b	s b	s b	s b	s b	s b	s b	s b	#	#
s_+ s_+	s_- s_-	s_- s_-	s b	s b	s b	s b	s b	s b	s b	s b	#	#

Fig. 2. Space-time diagram of a real-time computation of a one-way Mealy cellular automaton with strongly reversible cells on input a^3b^8.

Another related concept has been studied in [10]. Based on the observation, that in reversible one-way cellular automata information flow is from right to left in a forward computation and from left to right in a backward computation, a one-way cellular automaton is said to be reversible if there exists a reverse local transition function that computes the predecessor states. Due to the domain S^2

and the range S, obviously, the local transition function cannot be injective in general. However, since for reverse computation steps the flow of information is reversed as well, for the reverse transition function, each cell receives the state of its *left* neighbor. For example, let $s_1s_1s_2s_1$ be the states of four adjacent cells, and $\delta(s_1, s_1) = s_1$, $\delta(s_1, s_2) = s_2$, and $\delta(s_2, s_1) = s_1$, then the successor states of the three left cells are $s_1s_2s_1$. So, for the reverse transition function δ^R we obtain $\delta^R(s_1, s_2) = s_1$ and $\delta^R(s_2, s_1) = s_2$ and, thus, such a behavior is possible in reversible one-way cellular automata. However, the single cell is not a reversible finite automaton, since $\delta^{\leftarrow}(s_1, s_1) = s_1$ and $\delta^{\leftarrow}(s_1, s_1) = s_2$. On the other hand, let $s_1s_2s_2$ as well as $s_2s_3s_3$ be the states of three adjacent cells, and $\delta(s_1, s_2) = s_1$, $\delta(s_2, s_2) = s_2'$, $\delta(s_2, s_3) = s_1$, and $\delta(s_3, s_3) = s_2'$. Then the successor states of the two left cells are s_1s_2' in both cases. So, a reverse transition function δ^R cannot exist since it must map (s_1, s_2') to s_2 and to s_3. However, the transitions $\delta^{\leftarrow}(s_1, s_2) = s_1$, $\delta^{\leftarrow}(s_2', s_2) = s_2$, $\delta^{\leftarrow}(s_1, s_3) = s_2$, and $\delta^{\leftarrow}(s_2', s_3) = s_3$ do not violate the reversibility of the finite automaton used as single cell.

3   The Computational Capacity of One-Way Mealy Cellular Automata with Weakly Reversible Cells

Here, we explore the computational capacity of WRC-OMCA. To this end, we start to shed light on the role played by the sink states in such devices.

Lemma 2. *Let M be a WRC-OMCA whose cells are reversible disregarding sink states. Then an equivalent WRC-OMCA where all cells are reversible including sink states can effectively be constructed.*

Proof. Let $M = \langle S, F, A, B, \perp, \#, \delta \rangle$ be a WRC-OMCA whose cells are reversible except for sink states (see Fig. 3).

We consider the state graph of a cell of M. For every sink state s the following steps are repeated. Let G be the part of the graph that does neither include s nor any edge to s. The first step is to remove irreversibility for the edges that enter the sink state from some states in G. To this end, state s is copied as many times as there are incoming edges from states in G. Now these edges are directed to different copies of s.

The next step is to remove the irreversibility caused by the looping edges and the incoming edge from a state in G. To this end, a new copy $\bar{B} = \{\bar{b} \mid b \in B\}$ of B is used. Each edge from a copy of a sink state to itself labeled a, b is relabeled by $a, \bar{b}$. The state graph obtained so far is weakly reversible. However, by providing a copy of B the number of messages that may be sent to neighboring cells is increased. Since $A = B$, additional edges have to be included. To overcome this problem, for every edge in G labeled a, b, an additional edge between the same states labeled $\bar{a}, b$ is included in G'. Again, this step preserves weak reversibility. Altogether, we have constructed an equivalent WRC-OMCA, since every pair (a, b) from $B \times B$ induces an injective partial mapping from the state set S to itself.   □

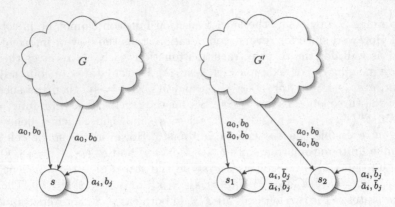

Fig. 3. How to make an automaton with sink state reversible. Automaton G is reversible except for the sink state s, where $0 \leq i, j \leq |B| - 1$ (left). Automaton G' is reversible including the two sink states s_1 and s_2 (right). Every edge in G labeled a, b is labeled a, b and $\bar{a}, b$ in G' which preserves weak reversibility.

The idea used to prove Lemma 2 can in fact be generalized. So, it turns out that even WRC-OMCAs have the full computational capacity of OMCAs.

Theorem 3. *Let M be an OMCA. Then an equivalent WRC-OMCA with all cells reversible including sink states can effectively be constructed.*

4   Reversible One-Way Partitioned Cellular Automata

Now we turn to discuss the details of the definitions in comparison with another related model. In [18] reversible two-way partitioned cellular automata are studied in terms of language recognition. The important concept of partitioned cellular automata is well-suited to define the notion of reversibility of cellular automata computations. In detail, the cells of a one-way *partitioned* cellular automaton have partitioned states that is, a state consists of a state part that represents the actual state and a message part the represents the message to be sent to the left neighbor. This message is created by the transition function during a transition. So, as for Mealy cellular automata the transition depends on the current state part and the current message part of its neighbor, and gives the new state part and the message part to be sent to the left, where initially each cell sends a message corresponding to its input symbol and the rightmost cell receives information associated with a boundary symbol on its free input line. So far, Mealy and partitioned cellular automata formalize similar concepts, but in partitioned cellular automata the message to be sent is a part of the state, while in Mealy cellular automata it is not. This makes a difference for reversibility considerations.

Formally, a *one-way partitioned cellular automaton* (OPCA) is a system $\langle S, F, A, \#, \delta \rangle$, where $S = T \times C$ is the finite, nonempty set of *cell states*, where T

is the *message part* and C is the *state part*, $F \subseteq S$ is the set of *accepting states*, $\# \in T$ is the distinguished *boundary message*, A is the nonempty set of *input symbols* with $A \subseteq C$ and $A \subseteq T$, and $\delta : C \times T \to T \times C$ is the *local transition function*.

Given some cell state $s = (t, c)$, the projection on its message part t is denoted by $\pi_T(s)$ and the projection on its state part c is denoted by $\pi_C(s)$.

A *configuration* of a one-way partitioned cellular automaton $\langle S, F, A, \#, \delta \rangle$ is a mapping $c : \{1, 2, \ldots, n\} \to S$, for $n \geq 1$, that assigns a state to each cell. The operation starts in a so-called *initial configuration*, which is defined by the given input $w = a_1 a_2 \cdots a_n \in A^+$. We set $c_0(i) = (a_i, a_i)$, for $1 \leq i \leq n$. Successor configurations are computed according to the global transition function Δ. Let c be a configuration with $n \geq 1$, then the successor configuration c' is

$$c' = \Delta(c) \iff \begin{cases} c'(i) = \delta(\pi_C(c(i)), \pi_T(c(i+1))), i \in \{1, 2, \ldots, n-1\} \\ c'(n) = \delta(\pi_C(c(n)), \#) \end{cases}.$$

A partitioned cellular automaton is said to be *(locally) reversible* (REV-OPCA) if and only if its local transition function is injective. So, given a state (s, m) and a transition $\delta(s, \ell) = (m', s')$, by the injectivity, from (m', s') the predecessor state s of the cell and the message ℓ received in the previous step are uniquely determined. The latter is part of the predecessor state of the right neighbor. In particular, the message part m of the cell cannot be determined. Instead, it is uniquely determined from the left neighbor. So, looking at the whole configuration, the predecessor configuration can be computed. However, the single cell is not necessarily a reversible finite automaton.

In RC-OMCAs the single cells have to be reversible OMCAs. So, for example, transitions $\delta(s_1, a_1) = (s, b)$ and $\delta(s_2, a_2) = (s, b)$ are allowed, where $\delta^-(s, a_1, b) = s_1$ and $\delta^-(s, a_2, b) = s_2$. These transitions are forbidden in reversible partitioned cellular automata since they violate the injectivity of δ.

The next theorem marks a lower bound for the computational capacity of real-time OPCAs. It says that a real-time OPCA is at least as powerful as a deterministic finite automaton (DFA), where a DFA is a DFMA with a singleton output alphabet, thus, the output is omitted from the transition function. Since it is well known that there are regular languages that are not accepted by reversible DFAs [2,20], the next theorem provides a construction of a reversible cellular device that simulates any possibly irreversible regular language.

Theorem 4. *Let L be a regular language. Then L is accepted by a real-time REV-OPCA.*

Proof. Since the regular languages are closed under reversal, we may assume that language L^R is accepted by some DFA M with state set S, input alphabet A, initial state s_0, set of accepting states F, and transition function $\delta : S \times A \to S$. Moreover, we may assume that the initial state s_0 is left with the very first transition and never reentered.

The idea for the simulation of M by a real-time OPCA M' is straightforward. The cells of M' run through a loop that keeps their initial states, while at the right end a signal is set up that moves through the array with maximum speed and computes and sends the states of the simulated REV-DFA M. A little extra attention has to be paid for implementing the local transition function of $M' = \langle S', F', A, \#, \delta' \rangle$ injectively. Depending on M we identify the boundary symbol $\#$ of M' with the initial state s_0 of M and define $\# = s_0$, $T = S \cup A$, $C = A \cup (S \times A)$, $F' = \{(s,a) \in S \times A \mid \delta(s,a) \in F\}$, and $\delta'(a,b) = (b,a)$, for all $a, b \in A$, $\delta'(a, \#) = (\delta(s_0, a), (s_0, a))$, for all $a \in A$, $\delta'(a, s) = (\delta(s,a), (s,a))$, for all $a \in A$ and $s \in S \setminus \{s_0\}$, and $\delta'((s,a), \#) = (\#, (s,a))$, for all $(s,a) \in S \times A$ (see Fig. 4).

a_1 a_1	a_2 a_2	a_3 a_3	a_4 a_4	a_5 a_5	$\#$
a_2 a_1	a_3 a_2	a_4 a_3	a_5 a_4	s_1 (s_0,a_5)	$\#$
a_3 a_1	a_4 a_2	a_5 a_3	s_2 (s_1,a_4)	$\#$ (s_0,a_5)	$\#$
a_4 a_1	a_5 a_2	s_3 (s_2,a_3)	$\#$ (s_1,a_4)	$\#$ (s_0,a_5)	$\#$
a_5 a_1	s_4 (s_3,a_2)	$\#$ (s_2,a_3)	$\#$ (s_1,a_4)	$\#$ (s_0,a_5)	$\#$
s_5 (s_4,a_1)	$\#$ (s_3,a_2)	$\#$ (s_2,a_3)	$\#$ (s_1,a_4)	$\#$ (s_0,a_5)	$\#$

Fig. 4. A real-time REV-OPCA accepting a regular language, where $s_{i+1} = \delta(s_i, a_{5-i})$, for $0 \leq i \leq 4$. The input is accepted if $\delta(s_4, a_1) = s_5$ is an accepting state in M.

An inspection of the transition function δ' and taking into account that the initial state of M is left in the very first transition and never reentered shows the injectivity of δ'. Let the input be $a_1 a_2 \cdots a_n$. At time step 1, the rightmost cell n initiates the signal by calculating and sending $s_1 = \delta(s_0, a_n)$. In general, for $0 \leq i \leq n-1$, the signal reaches cell $n-i$ at time $i+1$ and calculates and sends state $s_{i+1} = \delta(s_i, a_{n-i})$. The accepting states of M' are defined as those states sending an accepting state of M. So, M' accepts $L(M)^R = (L^R)^R = L$. $\square$

The next construction shows that real-time REV-OPCAs are also able to accept non-regular context-free languages.

Lemma 5. *There is a non-regular context-free language that is accepted by some real-time REV-OPCA.*

Proof. We use a language L over alphabet $\{a, b\}^*$ as witness that has the property $L \cap a^* b^* = \{a^m b^n \mid n \geq m \geq 1\}$. Since the regular languages are closed under intersection and $\{a^m b^n \mid n \geq m \geq 1\}$ is not regular, L is not regular either.

First, we partially construct a real-time REV-OPCA $M = \langle S, F, A, \#, \delta \rangle$ that accepts inputs from $\{a^m b^n \mid n \geq m \geq 1\}$. The basic idea is to send a signal with half speed from the rightmost a-cell to the left and a second signal that moves with maximum speed from the rightmost b-cell to the left. Whenever the second signal reaches a cell that already has seen the first signal, the cell enters the accepting state. The crucial point is to implement this behavior with an injective transition function. To this end, we set $A = \{a, b\}$, $T = \{\#\} \cup A$, $C = A \cup \{1, 2\}$, and $F = \{(\#, 1)\}$.

General transitions and transitions that implement the half-speed signal are

1. $\delta(a, a) = (a, a)$,
2. $\delta(b, b) = (b, b)$,

3. $\delta(a, b) = (a, 1)$, and
4. $\delta(1, b) = (b, 1)$.

The second signal is identified with the message #. It is implemented by

5. $\delta(b, \#) = (\#, b)$,   6. $\delta(a, \#) = (\#, a)$, and   7. $\delta(1, \#) = (\#, 1)$.

By inspection of the right-hand sides of the transition function it is evident that δ is injective so far.

In order to provide further transition rules for inputs not of the form $a^m b^n$, we extend δ injectively by

8. $\delta(b, a) = (b, 2)$,   9. $\delta(2, a) = (a, 2)$, and   10. $\delta(2, \#) = (\#, 2)$.

Now inputs of the form $b^m a^n$ are treated similarly as inputs of the form $a^m b^n$. However, in the b-cells now state parts 2 are used that are not part of accepting states.

Since the #-signal is transmitted further to the left, one may obtain further accepting computations if the inputs are appropriately extended to the left. However, every input of the form $b^m a^n$, for $m, n \geq 1$, is rejected, and every input of the form $a^m b^n$, for $n \geq m \geq 1$, is accepted. In particular, we derive that $L(M) \cap a^* b^* = \{a^m b^n \mid n \geq m \geq 1\}$, which shows the lemma. $\square$

5   Conclusions

We have introduced and discussed several notions of local reversibility for real-time OCAs. We have shown that weak reversibility can always be achieved, that is, every possibly irreversible real-time OCA computation can be realized by a real-time OCA composed of weakly reversible components. Concerning the other two notions, namely, strong reversibility and reversible partitioned OCAs we have the conjecture that both models are less powerful. However, both models are still able to accept complex languages such as the non-semilinear language given in Example 1 and the context-free, non-regular language used in Lemma 5. Apart from the question of whether both language classes can be separated from the general model, it would clearly be of interest to identify further language classes which can be accepted by these models. Finally, the strength of a model is to some extent documented by the undecidability of the usually investigated decidability questions such as emptiness, finiteness, inclusion, or equivalence. While all such questions are undecidable for real-time OCAs and hence also for real-time OCAs with weakly reversible cells, nothing is known yet on the status of the decidability questions for real-time OCAs with strongly reversible cells and real-time reversible partitioned OCAs.

Acknowledgments. We greatly acknowledge the valuable comments of the anonymous reviewers which, in particular, helped to improve the result of Theorem 4.

References

1. Abramsky, S.: A structural approach to reversible computation. Theor. Comput. Sci. **347**(3), 441–464 (2005)
2. Angluin, D.: Inference of reversible languages. J. ACM **29**(3), 741–765 (1982)
3. Bennett, C.H.: Logical reversibility of computation. IBM J. Res. Dev. **17**, 525–532 (1973)
4. Kari, J.: Theory of cellular automata: a survey. Theor. Comput. Sci. **334**(1–3), 3–33 (2005)
5. Kondacs, A., Watrous, J.: On the power of quantum finite state automata. In: Foundations of Computer Science (FOCS 1997), pp. 66–75. IEEE Computer Society (1997)
6. Kutrib, M., Malcher, A.: Fast reversible language recognition using cellular automata. Inform. Comput. **206**(9–10), 1142–1151 (2008)
7. Kutrib, M., Malcher, A.: Real-time reversible iterative arrays. Theor. Comput. Sci. **411**, 812–822 (2010)
8. Kutrib, M., Malcher, A.: Reversible pushdown automata. J. Comput. System Sci. **78**, 1814–1827 (2012)
9. Kutrib, M., Malcher, A.: One-way reversible multi-head finite automata. Theor. Comput. Sci. (to appear)
10. Kutrib, M., Malcher, A., Wendlandt, M.: Real-time reversible one-way cellular automata. In: Isokawa, T., Imai, K., Matsui, N., Peper, F., Umeo, H. (eds.) AUTOMATA 2014. LNCS, vol. 8996, pp. 56–69. Springer, Cham (2015). doi:10.1007/978-3-319-18812-6_5
11. Kutrib, M., Malcher, A., Wendlandt, M.: Reversible queue automata. Fund. Inform. **148**(3–4), 341–368 (2016)
12. Landauer, R.: Irreversibility and heat generation in the computing process. IBM J. Res. Dev. **5**, 183–191 (1961)
13. Lange, K.J., McKenzie, P., Tapp, A.: Reversible space equals deterministic space. J. Comput. System Sci. **60**, 354–367 (2000)
14. Morita, K.: Computation-universality of one-dimensional one-way reversible cellular automata. Inf. Process. Lett. **42**, 325–329 (1992)
15. Morita, K.: Reversible simulation of one-dimensional irreversible cellular automata. Theor. Comput. Sci. **148**(1), 157–163 (1995)
16. Morita, K.: Reversible computing and cellular automata - a survey. Theor. Comput. Sci. **395**, 101–131 (2008)
17. Morita, K.: Two-way reversible multi-head finite automata. Fund. Inform. **110**, 241–254 (2011)
18. Morita, K.: Language recognition by reversible partitioned cellular automata. In: Isokawa, T., Imai, K., Matsui, N., Peper, F., Umeo, H. (eds.) AUTOMATA 2014. LNCS, vol. 8996, pp. 106–120. Springer, Cham (2015). doi:10.1007/978-3-319-18812-6_9
19. Morita, K., Harao, M.: Computation universality of one dimensional reversible injective cellular automata. IEICE Trans. Inf. Syst. E **72**(1), 758–762 (1989)
20. Pin, J.-E.: On reversible automata. In: Simon, I. (ed.) LATIN 1992. LNCS, vol. 583, pp. 401–416. Springer, Heidelberg (1992). doi:10.1007/BFb0023844
21. Toffoli, T.: Computation and construction universality of reversible cellular automata. J. Comput. System Sci. **15**, 213–231 (1977)

Enumerating Orthogonal Latin Squares Generated by Bipermutive Cellular Automata

Luca Mariot[1,2]([✉]), Enrico Formenti[2], and Alberto Leporati[1]

[1] Dipartimento di Informatica, Sistemistica e Comunicazione,
Università degli Studi di Milano-Bicocca,
Viale Sarca 336, 20126 Milan, Italy
{luca.mariot,leporati}@disco.unimib.it
[2] Université Côte d'Azur, CNRS, I3S, Sophia Antipolis, France
mariot@i3s.unice.fr, enrico.formenti@unice.fr

Abstract. We consider the problem of enumerating pairs of bipermutive cellular automata (CA) which generate orthogonal Latin squares. Since the problem has already been settled for bipermutive CA with linear local rules, we address the general case of nonlinear rules, which could be interesting for cryptographic applications such as the design of cheater-immune secret sharing schemes. We first prove that two bipermutive CA generating orthogonal Latin squares must have pairwise balanced local rules. Then, we count the number of pairwise balanced bipermutive Boolean functions and enumerate those which generate orthogonal Latin squares up to $n = 6$ variables, classifying them with respect to their nonlinearity values.

Keywords: Cellular automata · Latin squares · Bipermutivity · Pairwise balancedness

1 Introduction

The construction of *orthogonal Latin squares* is a challenging combinatorial problem. Indeed, besides being one of the most researched topics in combinatorial design theory, orthogonal Latin squares also have numerous applications in cryptography, coding theory and the design of experiments [3,6,14].

Recently, a new construction of orthogonal Latin squares based on bipermutive cellular automata (CA) with linear local rules has been proposed in [10]. In particular, the authors proved that two linear bipermutive local rules generate a pair of orthogonal Latin squares if and only if their associated polynomials are relatively prime.

In this paper, we address the generalized problem of enumerating orthogonal Latin squares induced by *nonlinear* bipermutive CA, which could have interesting cryptographic applications. As a matter of fact, orthogonal Latin squares generated through nonlinear constructions can be employed in the design of *cheater-immune secret sharing schemes* [15].

© IFIP International Federation for Information Processing 2017
Published by Springer International Publishing AG 2017. All Rights Reserved
A. Dennunzio et al. (Eds.): AUTOMATA 2017, LNCS 10248, pp. 151–164, 2017.
DOI: 10.1007/978-3-319-58631-1_12

After covering in Sect. 2 the necessary preliminary notions about Latin squares and cellular automata, in Sect. 3 we first prove that the basic reversal and complementation operations on local rules preserve the orthogonality relation of the resulting Latin squares. Then, we show that two bipermutive local rules that give rise to orthogonal Latin squares must be *pairwise balanced*, which basically means that the four pairs $(0,0)$, $(1,0)$, $(0,1)$ and $(1,1)$ must occur an equal number of times in the superposition of their truth tables. Additionally, we prove that pairwise balancedness is a property preserved from the generating functions to the corresponding bipermutive rules, but not vice versa. In Sect. 4 we derive a formula for the number of pairwise balanced bipermutive rules, and apply a combinatorial algorithm to enumerate all those pairs which generate orthogonal Latin squares up to $n = 6$ variables. Finally, we classify these pairs with respect to their nonlinearity values. In Sect. 5 we sum up the contributions of this paper.

2   Preliminaries

In this section, we first recall the basic definitions about orthogonal Latin squares and cellular automata used throughout the paper. We then review the construction of orthogonal Latin squares based on linear bipermutive cellular automata described in [10].

2.1   Basic Definitions

In what follows, we denote by $[N]$ the set of the first N positive integer numbers, i.e. $[N] = \{1, \cdots, N\}$. We begin by defining the basic combinatorial objects of our interest, namely Latin squares:

Definition 1. *Let $N \in \mathbb{N}$. A* Latin square *of order N is a $N \times N$ matrix L such that each element of $[N]$ occurs exactly once in every row and in every column. Two Latin squares L_1 and L_2 of order N are called* orthogonal *if*

$$(L_1(i_1, j_1), L_2(i_1, j_1)) \neq (L_1(i_2, j_2), L_2(i_2, j_2)) \tag{1}$$

for all distinct pairs of coordinates $(i_1, j_1), (i_2, j_2) \in [N] \times [N]$.

Hence, two Latin squares are orthogonal if their *superposition* yields all the ordered pairs of the Cartesian product $[N] \times [N]$.

In this work, we consider a basic one-dimensional model of cellular automaton which can be considered as a special kind of vectorial Boolean function. For this reason, we first cover the necessary notions from the theory of cryptographic boolean functions, referring the reader to [1,2] for a more thorough presentation of the topic.

Let $\mathbb{F}_2$ and $\mathbb{F}_2^n$ respectively denote the finite field with two elements and the n-dimensional vector space over $\mathbb{F}_2$ (that is, the set of all binary n-tuples). In what follows, we assume that the 2^n vectors of $\mathbb{F}_2^n$ are lexicographically ordered, using least significant bit notation. A *Boolean function* of n variables is a mapping $f : \mathbb{F}_2^n \to \mathbb{F}_2$. The *truth table* of f is the vector $\Omega(f) \in \mathbb{F}_2^{2^n}$ defined as

$$\Omega(f) = (f(0,0,\cdots,0), f(1,0,\cdots,0),\cdots, f(1,1,\cdots,1)), \tag{2}$$

that is, $\Omega(f)$ specifies the output values of f for each of the possible 2^n values of the input vectors. Consequently, the set of all 2^n binary vectors coincides with the space of Boolean functions of n variables $\mathcal{F}_n$, which thus has size 2^{2^n}.

Let $f : \mathbb{F}_2^n \to \mathbb{F}_2$. The *reversal* $f_R : \mathbb{F}_2^n \to \mathbb{F}_2$ and the *complementation* $f_C : \mathbb{F}_2^n \to \mathbb{F}_2$ of f are defined as

$$f_R(x_1,\cdots,x_n) = f(x_R) = f(x_n,\cdots,x_1), \tag{3}$$
$$f_C(x_1,\cdots,x_n) = f(x_1,\cdots,x_n) \oplus 1, \tag{4}$$

for all $x = (x_1,\cdots,x_n) \in \mathbb{F}_2^n$. Clearly, both reversal and complementation are idempotent operations, i.e. $(f_R)_R = f$ and $(f_C)_C = f$.

A Boolean function $f : \mathbb{F}_2^n \to \mathbb{F}_2$ is called *affine* if it is defined as:

$$f(x_1,\cdots,x_n) = a \oplus a_1 \cdot x_1 \oplus \cdots \oplus a_n \cdot x_n \tag{5}$$

for all $x = (x_1,\cdots,x_n) \in \mathbb{F}_2^n$, where $a, a_1,\cdots,a_n \in \mathbb{F}_2$ and $\oplus$ and $\cdot$ respectively denote the XOR and AND operations. If $a = 0$, then the function is called *linear*.

The *nonlinearity* of a Boolean function $f : \mathbb{F}_2^n \to \mathbb{F}_2$ is defined as the minimum Hamming distance of f from the set of affine functions of n variables, a property which can be expressed using the Walsh transform of f. Given $f : \mathbb{F}_2^n \to \mathbb{F}_2$, the *Walsh transform* of f is the function $W_f : \mathbb{F}_2^n \to \mathbb{R}$ defined as

$$W_f(\omega) = \sum_{x \in \mathbb{F}_2^n} (-1)^{f(x) \oplus \omega \cdot x} \tag{6}$$

for all $\omega \in \mathbb{F}_2^n$, where $\omega \cdot x = \omega_1 x_1 \oplus \cdots \oplus \omega_n x_n$ is the *scalar product* between ω and x. The *spectral radius* of f, denoted as $W_{max}(f)$, is the maximum absolute value of its Walsh transform W_f over all vectors $\omega \in \mathbb{F}_2^n$. Then, the nonlinearity of f is formally defined as:

$$Nl(f) = 2^{n-1} - \frac{1}{2}W_{max}(f). \tag{7}$$

In this work, we focus mainly on CA based on bipermutive local rules. Formally, a *bipermutive* Boolean function is defined as follows:

Definition 2. *A boolean function $f : \mathbb{F}_2^n \to \mathbb{F}_2$ is called* bipermutive *if, by fixing either the leftmost or the rightmost $n-1$ input coordinates to any value $\tilde{x} \in \mathbb{F}_2^{n-1}$, the resulting restriction on the remaining coordinate is a permutation over $\mathbb{F}_2$. Equivalently, function $f : \mathbb{F}_2^n \to \mathbb{F}_2$ is bipermutive if there exists $\varphi : \mathbb{F}_2^{n-2} \to \mathbb{F}_2$ such that*

$$f(x_1, x_2, \cdots, x_{n-1}, x_n) = x_1 \oplus \varphi(x_2, \cdots, x_{n-1}) \oplus x_n \tag{8}$$

for all $x = (x_1, x_2, \cdots, x_{n-1}, x_n) \in \mathbb{F}_2^n$.

The function φ appearing in Eq. (8) is also called the *generating function* of f. Hence, the output of f is computed by XORing the leftmost and rightmost variables with the value of φ evaluated on the central variables. In [9], it has been shown that the nonlinearity of a bipermutive Boolean function is four times the nonlinearity of its generating function, i.e. $Nl(f) = 4 \cdot Nl(\varphi)$. Notice that a linear Boolean function is bipermutive if and only if its leftmost and rightmost coefficients a_1 and a_n are nonzero.

Vectorial Boolean functions generalize the concept of Boolean functions to multiple outputs. Given $n, m \in \mathbb{N}$, a *vectorial Boolean function* (or (n,m)-*function*) is a mapping $F : \mathbb{F}_2^n \to \mathbb{F}_2^m$. For all $i \in [m]$, the i-th *coordinate function* of F is the Boolean function $f_i : \mathbb{F}_2^n \to \mathbb{F}_2$ that specifies the i-th output bit of F, i.e. $f_i(x) = F(x)_i$ for all $x \in \mathbb{F}_2^n$.

Using the above notions on Boolean functions, we can now give a formal definition of cellular automaton.

Definition 3. *Let* $m, n \in \mathbb{N}$ *such that* $m \geq n$, *and let* $f : \mathbb{F}_2^n \to \mathbb{F}_2$ *be a Boolean function. A* one-dimensional cellular automaton *(CA) of length* m *with* local rule f *is a vectorial Boolean function* $F : \mathbb{F}_2^m \to \mathbb{F}_2^{m-n+1}$ *defined as*

$$F(x_1, \cdots, x_m) = (f(x_1, \cdots, x_n), \cdots, f(x_{m-n+1}, \cdots, x_m)) \qquad (9)$$

for all $x = (x_1, \cdots, x_m) \in \mathbb{F}_2^m$.

The local rule of a CA is usually identified by its *Wolfram code*, which is the decimal encoding of its truth table. On account of Definition 2, we call a CA *bipermutive* if its local rule is a bipermutive Boolean function.

A CA can be viewed as a vectorial Boolean function where each coordinate function f_i is the local rule f evaluated on the n input variables $x_i, \cdots, x_{i+n-1}$. From a different perspective, one can consider the input variables of the CA as *cells* whose state can be either 0 or 1, and where each of the first $m - n + 1$ cells updates in parallel its state by evaluating the local rule on the *neighborhood* formed by itself and the $n - 1$ cells to its right. Notice that the rightmost $n - 1$ input cells are not updated, hence there is no need to enforce any boundary condition. Remark also that, for the purposes of our work, we do not consider the *iterated behavior* of a CA produced by the repeated application of the local rule in successive time steps.

2.2   Latin Squares Generated by Cellular Automata

We now review the method for constructing Latin squares through bipermutive cellular automata, following the notation of [10]. Let us consider a CA $F : \mathbb{F}_2^{2(n-1)} \to \mathbb{F}_2^{n-1}$ based on a local rule $f : \mathbb{F}_2^n \to \mathbb{F}_2$ of n variables. Thus, F associates configurations of length $2(n - 1)$ to configurations of length $n - 1$. We can define a square matrix S_F by using the leftmost and rightmost $n - 1$ input variables of F to index respectively the rows and the columns of S_F, while the $n - 1$ output variables of F are employed to represent the entries of S_F at the respective input coordinates. More formally, let $N = 2^{n-1}$ and assume that

$\phi : \mathbb{F}_2^{n-1} \to [N]$ is a one-to-one mapping between $\mathbb{F}_2^{n-1}$ and $[N]$, and let ψ be the inverse mapping of ϕ. Then, the square associated to a CA of length $2(n-1)$ is defined as follows:

Definition 4. *Let $F : \mathbb{F}_2^{2(n-1)} \to \mathbb{F}_2^{n-1}$ be a CA with local rule $f : \mathbb{F}_2^n \to \mathbb{F}_2$. The* square *associated to F is the square matrix S_F of size $N \times N$ defined for all $1 \le i, j \le N$ as:*

$$S_F(i,j) = \phi(F(\psi(i)\|\psi(j))), \tag{10}$$

where $\psi(i)\|\psi(j) \in \mathbb{F}_2^{2(n-1)}$ is the concatenation of vectors $\psi(i), \psi(j) \in \mathbb{F}_2^{n-1}$.

We remark that this particular representation has been adopted in several works in the CA literature, even though under a different guise. Indeed, one can consider the square associated to a CA as the *Cayley table* of an algebraic structure $(A, \circ)$, where A is a set of size 2^{n-1} isomorphic to $\mathbb{F}_2^{n-1}$, and $\circ$ is a binary operation over A. The two operands $x, y \in A$ are represented by the vectors respectively composed of the leftmost and rightmost $n - 1$ input cells of the CA, while the $n - 1$ output cells represent the result $z = x \circ y$. To the best of our knowledge, the first who employed this algebraic characterization of cellular automata were Pedersen [13] and Eloranta [4], respectively for investigating periodicity and partial reversibility of CA. Other works in this line of research include Moore and Drisko [12], which studied the algebraic properties of the square representation of CA, and Moore [11], which considered the computational complexity of predicting CA whose local rules define solvable and nilpotent groups.

Depending on the underlying local rule, different algebraic structures can arise from the Cayley table of a CA. The case of *quasigroups* is especially interesting for the purposes of our work, since they are related to Latin squares. An algebraic structure $(Q, \circ)$ is a quasigroup if for all $x, y \in Q$ the two equations $x \circ z = y$ and $z \circ x = y$ have a unique solution for every $z \in Q$. When the support set Q is finite, the structure $(Q, \circ)$ is a quasigroup if and only if its Cayley table is a Latin square of order $|Q|$ [14].

A natural question to investigate is what classes of CA generate Latin squares (or equivalently, quasigroups). The following result shows that this is the case for bipermutive CA:

Lemma 1. *Let $F : \mathbb{F}_2^{2(n-1)} \to \mathbb{F}_2^{n-1}$ be a bipermutive CA with rule $f : \mathbb{F}_2^n \to \mathbb{F}_2$. Then, the square S_F induced by F is a Latin square of order $N = 2^{n-1}$.*

A proof of this fact which uses the characterization of quasigroups can be found in [4], while [10] reports a similar proof directly based on Latin squares.

Since bipermutive CA induce Latin squares, one could additionally investigate which pairs of them are orthogonal. This problem has been settled in [10] for the case of *linear* bipermutive CA. Considering Eq. (8), this means that the generating functions of the local rules are linear. More precisely, let $f, g : \mathbb{F}_2^n \to \mathbb{F}_2$ be bipermutive Boolean functions with linear generating functions $\varphi, \gamma : \mathbb{F}_2^{n-2} \to \mathbb{F}_2$ respectively defined as:

$$\varphi(x_2, \cdots, x_{n-1}) = a_2 x_2 \oplus \cdots \oplus a_{n-1} x_{n-1}, \tag{11}$$

$$\gamma(x_2, \cdots, x_{n-1}) = b_2 x_2 \oplus \cdots \oplus b_{n-1} x_{n-1}, \tag{12}$$

where $a_i, b_i \in \mathbb{F}_2$ for $i \in \{2, \cdots, n-1\}$. In this case, we can associate to f and g two polynomials $p_f(X), p_g(X) \in \mathbb{F}_2[X]$ of degree $n-1$ using the coefficients of their generating functions as follows:

$$p_f(X) = 1 + a_2 X \oplus \cdots \oplus a_{n-1} X^{n-2} + X^{n-1}, \tag{13}$$

$$p_g(X) = 1 + b_2 X \oplus \cdots \oplus b_{n-1} X^{n-1} + X^{n-1}. \tag{14}$$

The following result proved in [10] gives a necessary and sufficient condition on the polynomials p_f and p_g in order for F and G to generate orthogonal Latin squares:

Theorem 1. *Let $F, G : \mathbb{F}_2^{2(n-1)} \to \mathbb{F}_2^{n-1}$ be two bipermutive CA with linear local rules $f, g : \mathbb{F}_2^n \to \mathbb{F}_2$, and let p_f and p_g be their associated polynomials. Then, the Latin squares S_F and S_G respectively associated to F and G are orthogonal if and only if p_f and p_g are coprime.*

3   Main Results

Since the problem of characterizing pairs of bipermutive CA which generate orthogonal Latin squares has already been solved in [10] when the underlying local rules are linear, we now consider the more general case of nonlinear bipermutive CA. In order to tackle this problem, in this section we prove some results that allow us to reduce the search space of all bipermutive functions pairs. Then, we will use these results to enumerate all pairs of bipermutive CA that give rise to orthogonal Latin squares, with local rules of up to $n = 6$ variables.

Let $\mathcal{B}_n$ be the set of all pairs of bipermutive Boolean functions of n variables. As bipermutive functions are defined by their generating functions of $n-2$ variables, for all $n \geq 2$ it follows that $|\mathcal{B}_n| = |\mathcal{G}_n|$, where $\mathcal{G}_n = \{(\varphi, \gamma) \in \mathcal{F}_{n-2} \times \mathcal{F}_{n-2}\}$. Since $|\mathcal{F}_{n-2}| = 2^{2^{n-2}}$, the size of $\mathcal{G}_n$ is $2^{2^{n-2}} \cdot 2^{2^{n-2}} = 2^{2^{n-1}}$, meaning that $\mathcal{G}_n$ is isomorphic to $\mathcal{F}_{n-1}$, i.e. the set of Boolean functions of $n-1$ variables.

Clearly, if two bipermutive CA induced by a pair of local rules (f, g) give rise to orthogonal Latin squares, then the CA defined by the swapped pair (g, f) will generate the same orthogonal Latin squares in reverse order. We now show that the basic transformations of reversal and complementation introduced in Sect. 2.1 preserve the orthogonality relation as well:

Lemma 2. *Let $F, G : \mathbb{F}_2^{2(n-1)} \to \mathbb{F}_2^{n-1}$ be two bipermutive CA respectively defined by local rules $f, g : \mathbb{F}_2^n \to \mathbb{F}_2$ of n variables, and let S_F, S_G be the associated Latin squares of order 2^{n-1}. Additionally, let F_R, G_R and F_C, G_C be the CA respectively defined by the reverses f_R, g_R and the complements f_C, g_C of f, g, and let S_{F_R}, S_{G_R} and S_{F_C}, S_{G_C} be the corresponding Latin squares. Then, the following hold:*

– *S_F and S_G are orthogonal if and only if S_{F_R}, S_{G_R} are orthogonal.*
– *S_F and S_G are orthogonal if and only if S_{F_C}, S_{G_C} are orthogonal.*

Proof. Since both reversal and complementation are idempotent transformations, it suffices to show only one direction of the implications, i.e. assuming that S_F and S_G are orthogonal. This means that

$$(F(x||y), G(x||y)) \neq (F(x'||y'), G(x'||y'))$$

for all distinct pairs $(x,y), (x',y') \in \mathbb{F}_2^{n-1} \times \mathbb{F}_2^{n-1}$, since the mapping ϕ which associates binary vectors of length $n-1$ to positive integers in the range $\{1, \cdots, 2^{n-1}\}$ is bijective.

Let us now consider the CA F_R induced by the reversed local rule f_R. Then, for all $(x,y) \in \mathbb{F}_2^{n-1} \times \mathbb{F}_2^{n-1}$ with $x = (x_1, \cdots, x_{n-1})$ and $y = (y_1, \cdots, y_{n-1})$, it follows that

$$F_R(x||y) = (f_R(x_1, \cdots, x_{n-1}, y_1), \cdots, f_R(x_{n-1}, y_1, \cdots, y_{n-1})) =$$
$$= (f(y_1, x_{n-1}, \cdots, x_1), \cdots, f(y_{n-1} \cdots, y_1, x_{n-1})) = F(y_R||x_R)_R,$$

i.e., the output value of the reversed CA F_R is obtained by computing the reversed output of F evaluated on the reversed input $y_R||x_R$. Analogously, the same fact holds for G_R with respect to G. Since for all $(x,y), (x',y') \in \mathbb{F}_2^{n-1} \times \mathbb{F}_2^{n-1}$ such that $(x,y) \neq (x',y')$ one has that $(y_R, x_R) \neq (y_R', x_R')$, it follows that

$$(F(y_R||x_R)_R, G(y_R||x_R)_R) \neq (F(y_R'||x_R')_R, G(y_R'||x_R')_R),$$

which means that S_{F_R} and S_{G_R} are orthogonal Latin squares.

Next, let us consider the CA F_C induced by the complemented local rule f_C. The output value of F_C over $x||y$ is

$$F_C(x||y) = (f_c(x_1, \cdots, x_{n-1}, y_1), \cdots, f_c(x_{n-1}, y_1, \cdots, y_{n-1})) =$$
$$= (1 \oplus f(x_1, \cdots, x_{n-1}, y_1), \cdots, 1 \oplus f(x_{n-1}, y_1, \cdots, y_{n-1})) =$$
$$= \underline{1} \oplus F(x||y),$$

where $\underline{1} = (1, \cdots, 1) \in \mathbb{F}_2^{n-1}$. Similarly for G_C, one has $G_C(x||y) = \underline{1} \oplus G(x||y)$. Given two pairs $(x,y), (x',y') \in \mathbb{F}_2^{n-1} \times \mathbb{F}_2^{n-1}$ such that $(x,y) \neq (x',y')$, it clearly holds that $(\underline{1} \oplus x, \underline{1} \oplus y) \neq (\underline{1} \oplus x', \underline{1} \oplus y')$, from which it follows

$$(\underline{1} \oplus F(x||y), \underline{1} \oplus G(x||y)) \neq (\underline{1} \oplus F(x'||y'), \underline{1} \oplus G(x'||y')).$$

As a consequence, the Latin squares S_{F_C} and S_{G_C} are orthogonal. □

We now turn to analyze the truth tables of bipermutive rules whose CA generate orthogonal Latin squares. As an example, consider the pair of functions $f, g : \mathbb{F}_2^3 \to \mathbb{F}_2$ defined as $f(x_1, x_2, x_3) = x_1 \oplus x_3$ and $g(x_1, x_2, x_3) = x_1 \oplus x_2 \oplus x_3$, namely rules 90 and 150 using Wolfram's numbering convention. The Latin squares of order $N = 4$ induced by the corresponding bipermutive CA $F, G : \mathbb{F}_2^4 \to \mathbb{F}_2^2$ are orthogonal, since by Theorem 1 f and g are linear and their associated polynomials $p_f(X) = 1 + X^2$ and $p_g(X) = 1 + X + X^2$ are coprime. The truth tables $\Omega(f), \Omega(g) \in \mathbb{F}_2^8$ are the following:

$$\Omega(f) = (0, 1, 0, 1, 1, 0, 1, 0), \tag{15}$$
$$\Omega(g) = (0, 1, 1, 0, 1, 0, 0, 1). \tag{16}$$

Placing side by side these truth tables, one can see that there are $2^{3-2} = 2$ occurrences of each of the four pairs $(0,0)$, $(1,0)$, $(0,1)$ and $(1,1)$. We call this property *pairwise balancedness*, formally defined below:

Definition 5. *Two Boolean functions* $f, g : \mathbb{F}_2^n \to \mathbb{F}_2$ *of n variables are* pairwise balanced *if the* $(n,2)$-*function* $(f,g) : \mathbb{F}_2^n \to \mathbb{F}_2^2$ *defined as* $(f,g)(x) = (f(x), g(x))$ *is balanced, that is* $|(f,g)^{-1}(y_1, y_2)| = 2^{n-2}$ *for all* $(y_1, y_2) \in \mathbb{F}_2^2$.

We now prove that pairwise balancedness is a necessary condition for a pair of bipermutive local rules whose CA generate orthogonal Latin squares:

Lemma 3. *Let* $F, G : \mathbb{F}_2^{2(n-1)} \to \mathbb{F}_2^{n-1}$ *be bipermutive CA respectively induced by local rules* $f, g : \mathbb{F}_2^n \to \mathbb{F}_2$, *and suppose that the associated Latin squares* S_F, S_G *are orthogonal. Then,* f *and* g *are pairwise balanced.*

Proof. Let $H : \mathbb{F}_2^{n-1} \times \mathbb{F}_2^{n-1} \to \mathbb{F}_2^{n-1} \times \mathbb{F}_2^{n-1}$ *be the function defined as*

$$H(x,y) = (F(x||y), G(x||y)) \tag{17}$$

for all $(x,y) \in \mathbb{F}_2^{n-1} \times \mathbb{F}_2^{n-1}$. *Since* S_F *and* S_G *are orthogonal, it follows that* H *is bijective.*

Consider two vectors $c, d \in \mathbb{F}_2^{n-1}$ *and, without loss of generality, suppose that the first components of* c *and* d, *namely* $c_1, d_1 \in \mathbb{F}_2$, *are fixed. We want to compute the number of preimages* $(x_1, \cdots, x_{n-1}, y_1) \in \mathbb{F}_2^n$ *which map to* (c_1, d_1) *under* (f, g). *In order to do so, we evaluate the ratio* N/M, *where:*

- *N is the number of input pairs* $(x,y) \in \mathbb{F}_2^{n-1} \times \mathbb{F}_2^{n-1}$ *such that the first components of the respective output pairs* $H(x,y)$ *equal* (c_1, d_1).
- *M is the number of input pairs* $(x,y) \in \mathbb{F}_2^{n-1} \times \mathbb{F}_2^{n-1}$ *where* x *and the first component of* y *are fixed.*

In this way, we count the total number of preimages of H *which map to* (c_1, d_1) *and normalize it by the number of preimages where the first n components of* H *are fixed, thus determining the number of preimages of* (c_1, d_1) *under* (f, g).

As H *is bijective,* N *corresponds to the number of pairs of binary vectors of length* $n-1$ *where the first components are fixed, which are* $2^{n-2} \cdot 2^{n-2} = 2^{2(n-2)}$. *On the other hand* $M = 2^{n-2}$, *since we only have* $n - 2$ *free variables in the input configuration of the CA. Hence, it follows that* $|(f,g)^{-1}(y_1, y_2)| = N/M = 2^{2(n-2)}/2^{n-2} = 2^{n-2}$. $\square$

In the next Lemma, we show that pairwise balanced generating functions induce pairwise balanced bipermutive CA:

Lemma 4. *Let* $\varphi, \gamma : \mathbb{F}_2^{n-2} \to \mathbb{F}_2$ *be pairwise balanced functions of* $n - 2$ *variables, with* $n > 2$. *Then, the bipermutive rules* $f, g : \mathbb{F}_2^n \to \mathbb{F}_2$ *induced by* φ *and* γ *are pairwise balanced.*

Proof. Let $(y_1, y_2) \in \mathbb{F}_2^2$. One has that $|(\varphi, \gamma)^{-1}(y_1, y_2)| = 2^{n-4}$, *since φ and γ are balanced. Additionally, for all $\tilde{x} = (x_2, \cdots, x_{n-1}) \in (\varphi, \gamma)^{-1}(y_1, y_2)$, let $(x_1, \tilde{x}, x_n) = (x_1, x_2, \cdots, x_{n-1}, x_n)$. Then, by Eq. (8) it follows that $(0, \tilde{x}, 0) \in (f, g)^{-1}(y_1, y_2)$ and $(1, \tilde{x}, 1) \in (f, g)^{-1}(y_1, y_2)$. Similarly, for all vectors $\tilde{x} \in (\varphi, \gamma)^{-1}(\bar{y}_1, \bar{y}_2)$ where $\bar{y}_1 = 1 \oplus y_1$ and $\bar{y}_2 = 1 \oplus y_2$, it holds that $(1, \tilde{x}, 0) \in (f, g)^{-1}(y_1, y_2)$ and $(0, \tilde{x}, 1) \in (f, g)^{-1}(y_1, y_2)$. Since the fiber of (y_1, y_2) under (f, g) is given by*

$$(f,g)^{-1}(y_1,y_2) = \{(0, \tilde{x}, 0) : \tilde{x} \in (\varphi, \gamma)^{-1}(y_1, y_2)\} \cup$$
$$\cup \{(1, \tilde{x}, 0) : \tilde{x} \in (\varphi, \gamma)^{-1}(\bar{y}_1, \bar{y}_2)\} \cup$$
$$\cup \{(0, \tilde{x}, 1) : \tilde{x} \in (\varphi, \gamma)^{-1}(\bar{y}_1, \bar{y}_2)\} \cup$$
$$\cup \{(1, \tilde{x}, 1) : \tilde{x} \in (\varphi, \gamma)^{-1}(y_1, y_2)\} \tag{18}$$

and since the four sets in Eq. (18) are disjoint and have the same cardinality of $(\varphi, \gamma)^{-1}(y_1, y_2)$, we can finally conclude that

$$|(f,g)^{-1}(y_1,y_2)| = 4 \cdot |(\varphi, \gamma)^{-1}(y_1, y_2)| = 4 \cdot 2^{n-4} = 2^{n-2}. \tag{19}$$

□

Remark that the converse of Lemma 4 does not hold. As a matter of fact, already for $n = 4$ variables there exist several instances of bipermutive functions pairs which produce orthogonal Latin squares (and hence are pairwise balanced) but whose generating functions are not pairwise balanced. An example is given by the two following linear rules:

$$f(x_1, x_2, x_3, x_4) = 1 \oplus x_1 \oplus x_3 \oplus x_4,$$
$$g(x_1, x_2, x_3, x_4) = x_1 \oplus x_4.$$

The generating function of g in this case is the constant function defined as $\gamma(x) = 0$ for all $x \in \mathbb{F}_2^2$. Hence, the pairs $(0, 1)$ and $(1, 1)$ never occur when superimposing the truth tables of the two generating functions of f and g.

4   Enumeration of Pairwise Balanced Bipermutive Rules

In this section, we enumerate all bipermutive rules pairs generating orthogonal Latin squares up to $n = 6$ variables and we classify them according to their nonlinearity.

The space of pairs of pairwise balanced generating functions is easily characterizable from the combinatorial point of view. In fact, for $n > 2$, each pairwise balanced pair $\varphi, \gamma : \mathbb{F}_2^{n-2} \to \mathbb{F}_2$ can be represented by a string s of length 2^{n-2} over the alphabet $A = \{1, 2, 3, 4\}$, where each symbol in s corresponds to the decimal encoding of one of the possible four pairs $(0, 0)$, $(1, 0)$, $(0, 1)$ and $(1, 1)$ occurring in the superposition of the truth tables. Since φ and γ are pairwise balanced, the string s must be balanced as well, meaning that the number of occurrences of each of the four symbols of A must be 2^{n-4}. Hence, the number of pairwise balanced pairs of generating functions of $n-2$ variables equals

$$\#Bal\mathcal{G}_n = \binom{2^{n-2}}{2^{n-4}} \cdot \binom{3 \cdot 2^{n-4}}{2^{n-4}} \cdot \binom{2^{n-3}}{2^{n-4}}. \tag{20}$$

As a matter of fact, to construct a balanced quaternary string of length 2^{n-2} one has first to select the positions of the 2^{n-4} occurrences of the first symbol, which can be chosen in $\binom{2^{n-2}}{2^{n-4}}$ different ways. Next, the 2^{n-4} occurrences of the second symbol must be chosen among the $2^{n-2} - 2^{n-4} = 3 \cdot 2^{n-4}$ remaining positions, which can be done in $\binom{3 \cdot 2^{n-4}}{2^{n-4}}$ different ways. Finally, for the 2^{n-4} occurrences of the third symbol one has to choose among $2^{n-2} - 2 \cdot 2^{n-4} = 2^{n-3}$ remaining positions, corresponding to $\binom{2^{n-3}}{2^{n-4}}$ possible choices. At this point, the occurrences of the fourth symbols are fixed.

However, we saw at the end of Sect. 3 that pairwise balancedness is not a necessary condition on the generating functions to obtain pairwise balanced bipermutive rules. Consequently, by enumerating all balanced quaternary strings of length 2^{n-2} one only explores a subset of the space of pairwise balanced bipermutive rules of n variables, and thus in turn a subset of the space of bipermutive CA pairs generating orthogonal Latin squares of order 2^{n-1}.

We thus have to resort to a combinatorial characterization of pairwise balanced bipermutive functions. To this end, we adopt the *graph representation* of bipermutive rules, originally introduced in [8]. Given $n \in \mathbb{N}$, consider an undirected graph $G = (V, E)$ where $V = \mathbb{F}_2^n$. Two nodes $v_1, v_2 \in V$ are connected by an edge if and only if they differ either in their leftmost or rightmost coordinates, while they agree on the remaining ones. Thus, G is composed of 2^{n-2} connected components, and each connected component is composed of 4 nodes all having degree 2. A Boolean function $f : \mathbb{F}_2^n \to \mathbb{F}_2$ can be represented as a labeling function $l_f : V \to \{0,1\}$ on the nodes of G. If f is bipermutive, then the labels of adjacent nodes must differ, while the labels of two nodes separated by a path of length 2 must be equal.

Clearly, given a pair of bipermutive functions $f, g : \mathbb{F}_2^n \to \mathbb{F}_2$, we can still represent them on the graph as a labeling function $l_{f,g} : V \to \{0,1\}^2$ on the nodes, where the labels are pairs specifying the outputs of the two functions. Assume that f and g are pairwise balanced: then, each pair $(y_1, y_2) \in \mathbb{F}_2^2$ occurs 2^{n-2} times as a label on G. As an example, Fig. 1 depicts the graph representation of rule 90 and 150, which are pairwise balanced. Additionally, due to the property of different labels on adjacent nodes, it follows that exactly half of the connected components contain all $(0,0)$ and $(1,1)$ labels, while the remaining half contain all $(1,0)$ and $(0,1)$ labels. Since there are only two types of connected components with respect to the labels $((0,0)/(1,1)$ and $(1,0)/(0,1))$, it means that we can choose them in $\binom{2^{n-2}}{2^{n-3}}$ different ways. Moreover, let $C = \{v_1, v_2, v_3, v_4\}$ be a connected component where $(v_1, v_2), (v_1, v_3), (v_4, v_2), (v_4, v_3) \in E$, and assume that the labels on the nodes are either $(0,0)$ or $(1,1)$. Then, the two labels can be arranged in two different ways, namely $(l_{f,g}(v_1), l_{f,g}(v_4)) = (0,0)$ and $(l_{f,g}(v_2, v_3)) = (1,1)$ or $(l_{f,g}(v_1), l_{f,g}(v_4)) = (1,1)$ and $(l_{f,g}(v_2), l_{f,g}(v_3)) = (0,0)$.

In the same way, the labels on the nodes of a connected component of the type $(1, 0)/(0, 1)$ can be placed in two different ways. As a consequence, each of the $\binom{2^{n-2}}{2^{n-3}}$ ways for choosing the connected components with labels $(0, 0)/(1, 1)$ and $(1, 0)/(0, 1)$ gives rise to $2^{2^{n-3}} \cdot 2^{2^{n-3}} = 2^{2^{n-2}}$ pairwise balanced bipermutive functions. We have thus proved the following result:

Lemma 5. *The number of pairwise balanced pairs of bipermutive Boolean functions* $f, g : \mathbb{F}_2^n \to \mathbb{F}_2$ *of* n *variables is:*

$$\#Bal\mathcal{B}_n = \binom{2^{n-2}}{2^{n-3}} \cdot 2^{2^{n-2}}. \tag{21}$$

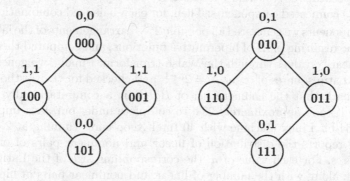

Fig. 1. Graph representation of the pairwise balanced bipermutive rules 90 and 150.

Table 1 reports the sizes of the search spaces for the sets of all pairs of bipermutive functions, the set of pairwise balanced generating functions and the set of pairwise balanced bipermutive functions of up to $n = 7$ variables.

One can notice that for $n \geq 7$ the resulting search space is too large to be exhaustively searched, even by focusing on the subsets of pairwise balanced generating functions. For this reason, we enumerated the set of pairwise balanced bipermutive functions $Bal\mathcal{B}_n$ only up to $n = 6$ variables. To this end, we implemented

Table 1. Sizes of the search spaces for the different types of sets of bipermutive functions pairs of up to $n = 7$ variables.

n	$\#\mathcal{B}_n$	$\#Bal\mathcal{G}_n$	$\#Bal\mathcal{B}_n$
3	16	0	8
4	256	24	96
5	65536	2520	17920
6	4294967296	63006300	843448320
7	$\approx 1.84 \cdot 10^{19}$	$\approx 9.96 \cdot 10^{15}$	$\approx 2.58 \cdot 10^{18}$

Table 2. Distribution of CA-based orthogonal Latin squares up to $n = 6$.

n	LS_size	#total	#linear	#nonlinear	nl_distribution
3	4×4	1	1	0	–
4	8×8	9	5	4	$(4, 4, 4)$
5	16×16	213	21	192	$(4, 4, 96)$, $(8, 8, 96)$, $(4, 4, 512)$, $(8, 8, 4020)$, $(12, 12, 17992)$
6	32×32	66685	85	66600	$(16, 16, 28388)$, $(20, 20, 14384)$, $(4, 12, 8)$, $(8, 16, 160)$, $(12, 20, 128)$, $(16, 24, 88)$

an algorithm by Knuth [7] to generate all balanced binary strings of length 2^{n-2}, where the positions set to 0 and 1 respectively correspond to the $(0, 0)/(1, 1)$ and $(1, 0)/(0, 1)$ connected components. Then, for each balanced combination of connected components we generated all possible $2^{2^{n-2}}$ arrangements of the labels, constructed the resulting pairs of bipermutive functions, and computed their respective nonlinearity values through the Walsh transform. Finally, we generated the associated Latin squares of order $N = 2^{n-1}$, and checked for their orthogonality.

We remark that the enumeration of $BalB_6$ is a computationally intensive task, since it took approximately 22 h to complete under our Java implementation on a 64-bit Linux machine with 40 Intel Xeon cores running at 2.4 GHz.

Table 2 reports the distribution of linear and nonlinear pairs of orthogonal Latin squares. For each value of n, the corresponding size of the Latin squares is reported, along with the number of linear and nonlinear pairs of bipermutive functions generating orthogonal Latin squares. Additionally, in the last column we report the distribution of nonlinearity values in triplets $(nl(f), nl(g), \#num)$ where $nl(f)$ and $nl(g)$ respectively denote the nonlinearity values of f and g, while $\#num$ is the number of pairs generating orthogonal Latin squares that achieve those values. Notice that all reported numbers are divided by 8, since we have to take into account the pairs with swapped order, which halve the resulting sets, and the reversal and complementation transformations, which by Lemma 2 additionally reduce them to a quarter.

As a qualitative remark on the distributions reported in Table 2, one may observe that linear pairs become more sparse as the number of variables n increases, while the majority of the pairs are nonlinear. Moreover, one can see that for $n = 6$ there are pairs with functions of different nonlinearities. This finding falsified our initial belief that two bipermutive functions inducing orthogonal Latin squares must have the same value of nonlinearity, an empirical observation which held up to $n = 5$ variables.

5   Conclusions

In this work, we considered the problem of exhaustively enumerating pairs of orthogonal Latin squares generated through bipermutive CA. We first proved that all pairs of bipermutive rules inducing orthogonal Latin squares must be pairwise balanced, meaning that the superposition of their truth tables must yield an equal number of occurrences of the four pairs $(0,0)$, $(1,0)$, $(0,1)$ and $(1,1)$. We then used a combinatorial algorithm to enumerate all pairwise balanced Boolean functions of up to $n = 6$ variables, finding those which generate orthogonal Latin squares and classifying them with respect to their nonlinearity values. The results of our computer search showed that, as the number of variables of the local rules increases, most of the orthogonal pairs are nonlinear. This could have interesting applications from the cryptographic point of view, since as mentioned in the Introduction orthogonal Latin squares arising from nonlinear constructions have relevance in the design of cheater-immune secret sharing schemes. We plan to study this issue in future research, in particular by investigating sufficient conditions that two nonlinear bipermutive CA must satisfy in order to generate orthogonal Latin squares. Another direction worth investigating is to analyze the pairs of nonlinear rules found in this paper from the perspective of *pseudorandom number generation*, and compare them with others stemming from different classifications, like those presented in [5,9].

References

1. Carlet, C.: Boolean functions for cryptography and error correcting codes. In: Crama, Y., Hammer, P.L. (eds.) Boolean Models and Methods in Mathematics, Computer Science, and Engineering, 1st edn., pp. 257–397. Cambridge University Press, New York (2010)
2. Carlet, C.: Vectorial Boolean functions for cryptography. In: Crama, Y., Hammer, P.L. (eds.) Boolean Models and Methods in Mathematics, Computer Science, and Engineering, 1st edn., pp. 398–469. Cambridge University Press, New York (2010)
3. Colbourn, C.J., Dinitz, J.H.: Making the mols table. In: Wallis, W.D. (ed.) Computational and Constructive Design Theory, pp. 67–134. Springer, Dordrecht (1996)
4. Eloranta, K.: Partially permutive cellular automata. Nonlinearity **6**(6), 1009–1023 (1993)
5. Formenti, E., Imai, K., Martin, B., Yunès, J.-B.: Advances on random sequence generation by uniform cellular automata. In: Calude, C.S., Freivalds, R., Kazuo, I. (eds.) Computing with New Resources. LNCS, vol. 8808, pp. 56–70. Springer, Cham (2014). doi:10.1007/978-3-319-13350-8_5
6. Keedwell, A.D., Dénes, J.: Latin Squares and Their Applications. Elsevier (2015)
7. Knuth, D.: The Art of Computer Programming, vol. 4, pre-fascicle 3a (2011)
8. Leporati, A., Mariot, L.: 1-resiliency of bipermutive cellular automata rules. In: Kari, J., Kutrib, M., Malcher, A. (eds.) AUTOMATA 2013. LNCS, vol. 8155, pp. 110–123. Springer, Heidelberg (2013). doi:10.1007/978-3-642-40867-0_8
9. Leporati, A., Mariot, L.: Cryptographic properties of bipermutive cellular automata rules. J. Cell. Autom. **9**(5–6), 437–475 (2014)
10. Mariot, L., Formenti, E., Leporati, A.: Constructing orthogonal latin squares from linear cellular automata. CoRR abs/1610.00139 (2016)
11. Moore, C.: Predicting nonlinear cellular automata quickly by decomposing them into linear ones. Phys. D Nonlinear Phenom. **111**(1–4), 27–41 (1998)

12. Moore, C., Drisko, A.A., et al.: Algebraic properties of the block transformation on cellular automata. Complex Syst. **10**(3), 185–194 (1996)
13. Pedersen, J.: Cellular automata as algebraic systems. Complex Syst. **6**(3), 237–250 (1992)
14. Stinson, D.R.: Combinatorial Designs - Constructions and Analysis. Springer, New York (2004)
15. Tompa, M., Woll, H.: How to share a secret with cheaters. J. Cryptol. **1**(2), 133–138 (1988)

Filling Curves Constructed
in Cellular Automata with Aperiodic Tiling

Gaétan Richard[✉]

Normandie Univ, UNICAEN, ENSICAEN, CNRS, GREYC, 14000 Caen, France
gaetan.richard@unicaen.fr

Abstract. In many constructions on cellular automata, information is transmitted with signals propagating through a defined background. In this paper, we investigate the possibility of using aperiodic tiling inside zones delimited by signals. More precisely, we study curves delineated by CA-constructible functions and prove that most of them can be filled with the NW-deterministic tile set defined by Kari [1]. The achieved results also hint a new possible way to study deterministic tile sets.

1 Introduction

Cellular automata are studied as a computation model with a theoretical computer science approach [2], and as a dynamical complex system with a multi-disciplinary approach [3]. The former approach uses specific portions of the whole configuration to encode computation and manages to achieve strong results on complexity or decidability. The latter looks at all configurations and try to prove results on the global behaviour. Roughly speaking, most results for computation deal with the existence of a specific configuration, whereas dynamics study the evolution for all of them.

It is a matter of fact there exists a few results that exploit the computational tools to prove behaviours on all configurations. The key example of such a result is the undecidability of nilpotency by Kari [1] in which he uses an aperiodic tiling to recursively embed everywhere in the configuration some parts of computation of arbitrary size. Using this tool, he can send ill-formed configurations "into a sink-hole" and work on the correct ones. More recently, a result has been obtained in the context of synchronisation using the same aperiodic tiling [4]. This construction generates linear zones embedding the aperiodic tiling is used.

In this paper, we investigate the possibility of constructing zones of various shapes embedding aperiodic tiling. Intuitively, we wonder whether we can fill inside of curves generated by methods similar to those used in [5,6] with Kari's tiling. The results also open interesting questions on determinism of tile sets.

The paper is organised as follow: in Sect. 2, we present the two used models (cellular automata and tile sets) along with notions of constructing and filling. In Sect. 3, we focus on determinism in tiling and look with more details at Kari's tiling. In Sect. 4, we state several results about aperiodic filling. At last, we propose some perspectives in Sect. 5.

© IFIP International Federation for Information Processing 2017
Published by Springer International Publishing AG 2017. All Rights Reserved
A. Dennunzio et al. (Eds.): AUTOMATA 2017, LNCS 10248, pp. 165–175, 2017.
DOI: 10.1007/978-3-319-58631-1_13

2   Constructions and Fillings

2.1   Cellular Automata

In this paper, we consider cellular automata in dimension one with nearest neighbours. Therefore, a *cellular automaton* is a pair (Q, δ) where Q is a finite set of *states* and $\delta : Q^3 \to Q$ is the *local transition function*. We assume that the automaton has a *quiescent state* $B \in Q$ satisfying $\delta(B, B, B) = B$. The global dynamics is achieved by applying uniformly and synchronously the local transition function to a bi-infinite line of cells: a *space-time diagram* D is an element of $S^{\mathbb{Z} \times \mathbb{N}}$ satisfying for all $(i, j) \in \mathbb{Z} \times \mathbb{N}, D(i, j+1) = \delta\left(D(i-1, j), D(i, j), D(i+1, j)\right)$. One can note that the space-time diagram is fully determined by the initial line $\{D(i, 0) \mid i \in \mathbb{Z}\}$. In this paper, we shall look at the space-time diagram D_q generated by an initial state $q \in Q$, defined as $D_q(0, 0) = q$ and $D_q(i, 0) = B$ for $i \neq 0$. By the previous remark, this is sufficient to define the whole space-time diagram.

We want to look at functions which can be constructed by cellular automaton as depicted in Fig. 1: in the space time diagram generated from a configuration containing only one cell in q and all other cells blank, the curve f is drawn with a subset of *marked* states $M \subset Q$. The portion above the curve is filled with non-marked non-blank symbols whereas the portion below the curve is filled with blanks.

Definition 1. *A function* $f : \mathbb{N} \to \mathbb{N}$ *is* constructible *if there exists a cellular automaton* $(Q \cup \{B\}, \delta)$, $q \in Q$, *and* $M \subset Q$, *such that* D_q *satisfies:*

- $D_q(i, j) = B$ *for* $i < 0$ *and* $j \in \mathbb{N}$;
- $D_q(i, f(i)) \in M$ *for* $i > 0$;
- $D_q(i, j) = B$ *for* $i > 0$ *and* $j < f(i)$;
- $D_q(i, j) \in Q \backslash M$ *for* $i \in \mathbb{N}$ *and* $j > f(i)$.

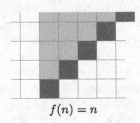

$$f(n) = n$$

Fig. 1. A basic example of constructible function

It can be seen that the definition only makes sense for function $f : \mathbb{N} \to \mathbb{N}$ that are strictly increasing and satisfy $f(0) = 0$.

This notion of constructibility has already been investigated, leading to hierarchy results [7] or even impossibility results [6]. Here, we will focus on filling the zone above the curve with an aperiodic tiling.

2.2   Tile Set

A *Wang tile t* is a square tile with coloured edges, as represented in Fig. 2. Formally, it is given by a quadruplet (t_e, t_w, t_n, t_s) of symbols chosen among a finite alphabet. A *tile set* τ is a finite set of Wang tiles. A *tiling* of the plane by τ is a map $T : \mathbb{Z}^2 \to \tau$ from the discrete plane to the tile set so that two tiles that share a common edge agree on the colour: For all integers i, j we have $T(i,j)_e = T(i+1,j)_w$ and $T(i,j)_n = T(i,j+1)_s$. A tiling T is said *periodic* if there exists $p, p' > 0$ such that for any $i, j \in \mathbb{Z}$, $T(i,j) = T(i+p,j) = T(i,j+p')$.

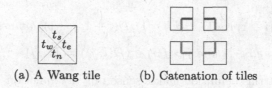

(a) A Wang tile   (b) Catenation of tiles

Fig. 2. Wang tiles

Given a constructible function, we want to see in which case the portion above the curve can be (in a constructive way) filled with tiles from a chosen tile set.

Definition 2. *A function f is fillable by a tile set τ if it is constructible by a cellular automaton $(Q \cup \{B\}, \delta)$ and there exists a projection $\pi : Q \to \tau$ and a tiling T such that for any $i \in \mathbb{N}$ and $j > f(i)$, $\pi(D_q(i,j)) = T(i,j)$.*

With this definition, it is not so difficult to prove that any constructible function can be filled by a tile set which has a periodic tiling.

Proposition 1. *Let τ be a tile set which admits a periodic tiling, then any constructible function f is fillable with τ.*

Proof. Let us take a tile set τ and $T : \mathbb{Z} \to \tau$, $T(i[p], j[p']) = \tau_{(i,j)}$ be one of its periodic tiling. Let f be a constructible function by a cellular automaton $(Q \cup \{B\}, \delta)$. We extend this automaton by adding a new layer containing the tiling $((Q \cup \{B\}, \tau \cup \{C\}), \delta')$ where

$$\delta'((q_l, t_l), (q_c, t_c), (q_r, t_r)) = (\delta(q_l, q_c, q_r), t(\delta(q_l, q_c, q_r), t_l, t_c))$$

and

$$t(q, t_l, t_c) = \begin{cases} \tau_{(i,j+1[p'])} & \text{if } q \in Q \backslash M \text{ and } t_c = \tau_{(i,j)} \\ \tau_{(i+1[p],j+1[p'])} & \text{if } q \in M \text{ and } t_l = \tau_{(i,j)} \\ C & \text{otherwise} \end{cases}$$

If we set $B' = (B, C)$ and $q'_0 = (q_0, \tau_{(0,0)})$ then it is easy to show by recurrence over j and i that the cellular automaton fills f. $\square$

This notion of fillable can be extended to get rid of the constructible constraint. This extension can model the fact that the curve dominated by an external action.

Definition 3. *A function f is* loosely fillable *with a tile set τ if there exists, a cellular automaton $(Q \cup \{B\}, \delta)$, $q \in Q$, a tiling T, a diagram $\tilde{D} \in \mathbb{Z} \times \mathbb{N} \to Q$, and a projection $\pi : Q \to \tau$ such that,*
*   if we define the diagram $\tilde{D} \in \mathbb{Z} \times \mathbb{N} \to Q$ as*

- $\tilde{D}(i,j) = B$ *for $i < 0$ and $j \in \mathbb{N}$;*
- $\tilde{D}(i,j) = B$ *for any $i \geq 0$ and $j < f(i)$;*
- $\tilde{D}(0,0) = q$;
- $\tilde{D}(i,j) = \delta(\tilde{D}(i-1,j-1), \tilde{D}(i,j-1), \tilde{D}(i+1,j-1))$ *otherwise,*

*   we have for any $i \in \mathbb{N}$ and $j > f(i)$, $\pi(\tilde{D}(i,j)) = T(i,j)$*

As previously, this definition makes sense only for functions $f : \mathbb{N} \to \mathbb{N}$ that are strictly increasing and satisfy $f(0) = 0$.

Remark that even if this definition removes the set M to mark the points $(i, f(i))$, it is possible to construct an automaton having this additional property since those points are the first ones in the column which are not in state B. The last remark implies that the construction of Proposition 1 applies as well to this extension.

Proposition 2. *Let τ be a tile set which admits a periodic tiling, then any function f is loosely fillable with τ.*

Proof. Just replace the test $q \in M$ by $q \neq B$ and $t_c = C$ in the previous construction.

3   Aperiodic Tile Set

Since filling is easy when the tile set admits a periodic tiling, let us look at what happens in the other case: A tile set τ is said to be *aperiodic* if there exists at least a valid tiling of the plane but no periodic tiling. A tile set of this kind was first found by Berger [8] to prove undecidability of the AEA fragment of logic. Since, several other aperiodic tile sets have been found [9–12].

3.1   From Tile Set Determinism to Cellular Automaton

In our case, the filling by cellular automata introduces some kind of determinism in the tiling. The corresponding notion has been introduced by Jarkko Kari: a tile set is said to be *north-west deterministic* if there is at most one possible tile (t_e, t_w, t_n, t_s) for every possible choice of (t_e, t_s). To keep the original terminology, north is going down.

This definition introduces one difficulty: the determinism has defined previously does not go in the same direction than the one induced by cellular

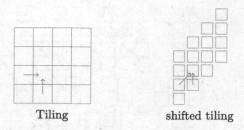

Tiling   shifted tiling

Fig. 3. From tile set determinism to cellular automata

automata. To overcome this problem, we shift each column of 1 relative to the previous one as depicted in Fig. 3.

With this operation, for any $x, y \in \mathbb{N}$, the point at coordinates (x, y) is mapped to the position $(i, j) = (x, y + x)$. Since we have taken f strictly increasing, $f(n) \geq n$ and thus those points are the only ones used in the definition of filling.

We slightly alter the definition of fillable to match this transformation:

Definition 4. *A function f is* fillable *by a north-west deterministic tile set τ if it is constructible by a cellular automaton $(Q \cup \{B\}, \delta)$ and there exists a projection $\pi : Q \to \tau$ and a tiling T such that for any $i \in \mathbb{N}$ and $j > f(i)$, $\pi(D_q(i, j)) = T(i, j - i)$.*

The loose fillable variant can be adapted in the same way.

This transformation achieves one easy implication from tile set determinism to cellular automata determinism: if a tile set is north-west deterministic, it can be "simulated" by a cellular automata. The converse is also true: any cellular automata can be "simulated" by a north-west deterministic tiling.

3.2   A NW-Deterministic Aperiodic Tile Set

This part is devoted to present the north-west deterministic tile set constructed by Kari in [1] which is built over the tiling from Robinson [9]). The reader can refer to those papers for the formal proof of aperiodicity.

This tile set can be described with two layers (corresponding to a Cartesian product). The first one is just a regular 4 coloured grid. The second one is usually depicted using lines. The tiles are depicted in Fig. 4.

Theorem 1 (Kari [1]). *The tile set depicted in Fig. 4 is north-west deterministic and aperiodic.*

Proof (Idea). The north-west deterministic property is easy to check. The aperiodicity is more complex but relies on the following fact: the first layer forces a 4-periodic background supporting squares of size 2 on the second layer. The second layer is designed such that squares of size n force the existence of squares of size $2n$.   □

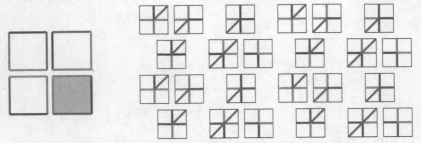

(a) Background (b) Signals (only light tiles may appear with starred background)

Fig. 4. A north-west deterministic aperiodic tile set

We are particularly interested in the specific tiling depicted in Fig. 5. Due to the auto-similar characteristic of the tile set, we can see (as represented in Fig. 6) that the tiling can be extended to the infinite quarter of plan. Also note that we can even construct a full tiling of the plan but this will not be of use here. The main useful point of this partial tiling is that its regularity will allow to achieve filling.

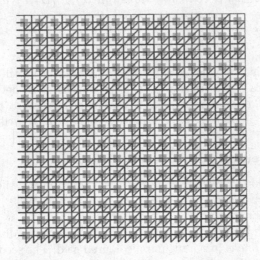

Fig. 5. A valid quarter of the plan tiling

4   Aperiodic Filling

In this section, we shall fill curves using Kari's tile set τ_K presented in the previous section and especially the portion of the tiling depicted in Fig. 5. To better see what happens in this tiling, Fig. 6 decorates the lines with different widths (that are not present in the tile set) to highlight the hierarchical structure in which squares assemble themselves into larger squares.

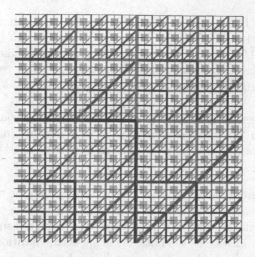

Fig. 6. Hierarchical vision of Kari's tiling

Since the tile set is NW deterministic, we define as $K : \tau_K \times \tau_K \to (\tau_K \cup \emptyset)$ the application which associate to two tiles, the unique one fitting in the NW corner (if it exists).

4.1   Linear Functions

The first result concerns identity function and corresponds to the case used in [4].

Proposition 3. *The function $f(n) = n$ is fillable by Kari's tile set.*

Proof. The constructibility of f is trivial.

To fill the curve, we shall add a layer similarly than in the proof of Proposition 1. This layer correspond to the tiling depicted previously. Inside the curve, we use NW determinism to guess the tile; On the borders (vertical on the left and diagonal on the right), we use the property that this correspond to lines over which the tiling has periodic behaviour of period 2 (see Fig. 6. Let us call $\tau_0^l = T(0, 2i), i \geq 0$, $\tau_1^l = T(0, 2i+1), i \geq 0, \tau_0^r = T(2i, 0), i > 0$, and $\tau_1^r = T(2i+1, 0), i \geq 0$.

Let $f(n) = n$ be constructible by a cellular automaton $(Q \cup \{B\}, \delta)$. We extend this automata by adding a new layer containing the tiling $((Q \cup \{B\}, \tau_K \cup \{C\}), \delta')$ where $\delta'((q_l, t_l), (q_c, t_c), (q_r, t_r)) = (\delta(q_l, q_c, q_r), t(\delta(q_l, q_c, q_r), t_l, t_c))$ and

$$
t(q, t_l, t_c) = \begin{cases} K(t_l, t_c) & \text{if } q \in Q \backslash M, t_l \neq C \text{ and } K(t_l, t_c) \neq \emptyset \\ \tau_{i+1[2]}^r & \text{if } q \in M \text{ and } t_l = \tau_i^r \\ \tau_1^r & \text{if } q \in M \text{ and } t_l = \tau_0^l \\ \tau_{i+1[2]}^l & \text{if } q \in Q \backslash M, t_l = C \text{ and } t_c = \tau_i^l \\ C & \text{otherwise} \end{cases}
$$

If we set $B' = (B, C)$ and $q'_0 = (q_0, T(0, 0) = \tau^l_0)$, we can see that the forth case of t correspond to cells which are on the Y-axis of both the tiling and the space-time diagram; the second case correspond to the X-axis of the tiling and the marked diagonal of the space-time diagram (the third case is the specific case happening at $(0, 0)$. At last, the first case cover the inside of the tiling and the curve.

As we have chosen a valid tiling, it can be noted that the condition $K(t_l, t_c) \neq \emptyset$ is always valid in the first case and thus C only occurs outside the curve. It follows that our conditions do correctly match boundaries and fill it with correct tiles.   □

The previous proof relies on the fact that the tiling we use has two non-collinear lines on which it is periodic. It may be interesting to search whether this condition is sufficient or necessary to achieve the result and eventually look at aperiodic deterministic tile set not exhibiting such condition if they exists.

The previous result can be extended to linear functions for which the coefficient is a rational number.

Corollary 1. *For any* $\alpha \in \mathbb{Q}$, $\alpha > 1$ *The function* $f(n) = \alpha n$ *is fillable by an aperiodic tile set.*

Proof. Constructing f consist on a signal of slope $1/\alpha$ which is the most simple one to construct.

To fill, it is sufficient to apply a grouping by bloc of $(n \times m)$ where $\alpha = m/n$ to Kari's tiling and reuse the previous proof. The grouping preserves the NW-determinism and the regular lines in the tiling are shifted to the correct place.
                                                                                                □

4.2   Over-Logarithmic Growth Functions

For now, we haven't still make use of the regularity visible in the tiling. This will allow us to gave a more general result:

Proposition 4. *Let* f *be a constructible function such that, for all* $n \in \mathbb{N}$, $f(n + 1) - f(n) > 2 \log_2(n)$, *then* f *is fillable with Kari's tile set.*

Proof. Once again, we shall use the same method as in the proof of Proposition 3. In fact, the only difference is in the second and third case in the definition of t (the method to find the correct tile on the curve). All other cases are unchanged. Thus, the only case posing problem is to find the tile to put at positions marked $(i, f(i))$.

Let us look at the corresponding part in the tiling (see Fig. 7). It can be easily seen that among the information needed, most of it can be deduced from the cell to the bottom left of it (the east one). In fact, the background colour, the diagonal and the horizontal signals can be deduced leaving only the vertical signal.

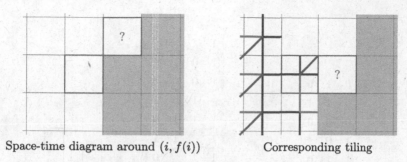

Space-time diagram around $(i, f(i))$ Corresponding tiling

Fig. 7. Guessing the correct tile at position $(i, f(i))$

Looking more in detail at Fig. 6, we can see that this signal has a high regularity: if we consider the x-th vertical line, the signal is periodic of period 2^{i+2} where i is the least significant non-zero bit of $x+1$ (that is, all odd columns have period 4, all even but not multiple of 4 have period 8, ...). More specifically, the signal alternates every 2^i on a cycle of 4 colours with the same starting point on the X-axis. Thus, to find the last missing piece of information, it is sufficient to keep track of $x+1$ and y (Y-value) corresponding to the position in the tiling and extract the correct bits.

Let us now look if it is possible to compute such values inside the interval. To do this, it is possible to maintain two integers recording x and y. To increment the value of y at each time steps, we need to use an increment with a lazy carry as in Fig. 8a. In this setup, the value of the counter can be read diagonally. It is also possible to shift the counter to the right keeping the same value as in Fig. 8b. The counter x is done in a similar way. Once both counters are present, the value needed can be easily extracted in $2\log(n)$ steps as depicted in Fig. 8c.

The previous method is correct but gives the result with a $2\log(n)$ delay (which explain the condition on the growth of f). However, since there are only 4 possibilities, it is easy to try all of them simultaneously and erase the incorrect one when the result arrives. To be fully coherent with our definition, which require the projection to know exactly which is the tile, the previous method can be slightly altered to guess one vertical line in advance and keep it in addition to the current tile.

One last point to underline is that our proof assume that there is always enough place to keep the counters. If it is true for x, it may not be the case for y. However, in the latter case, it means that $y = O(2^x)$ and since the vertical strip of width x the tiling is periodic of period 4^x, we can forget the high bits of y and still fill with a valid tiling. □

Even if our proof only use the marked M subset from the constructible function, this set can be emulated also in loose fill which allows to extend the result to this case.

Corollary 2. *Let f be a function such that, for all $n \in \mathbb{N}$, $f(n+1) - f(n) > 2\log_2(n)$, then f is loosely fillable with Kari's tile set.*

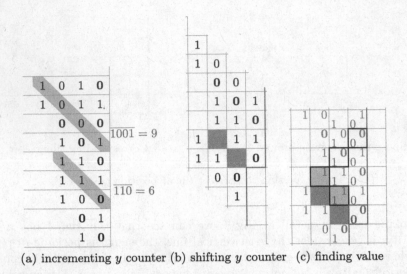

(a) incrementing y counter (b) shifting y counter (c) finding value

Fig. 8. Finding missing information for a new tile

For the constructibility, there exists a gap between $n - \log(n)$ and n where any function is not constructible [6]. At the moment, our result would suggest that there may be a similar gap for filling. This kind of result could prove interesting to better understand computation by cellular automata.

5   Conclusions and Perspectives

In this paper, we have shown that Kari's tile set can be used deterministically to fill curves. This method may open new possibilities to results on the whole set of configuration using computability tricks since it is easy to force this tile set to embed computation.

With respect to this specific tile set, the question of optimality of results in Sect. 4 is still an open question. The logarithmic growth condition is due to the maximal complexity of our method to find the next valid tile. It can be noted that the mean complexity is constant so there is perhaps room for improvement even using our proof.

All those results are based on the "regularity" of one tiling in periodic or Kari's tile set. This leads directly to try and generalise this for a wider range of tile set. In particular, one question is to find if there exists tile set for which the set of fillable function is different and understand why.

In this way, one class of tile set which may prove interesting are the one that can be depicted using substitutions. One first look at this set was done by Fernique and Ollinger [13]. The introduction of determinism and fill may allow to further discriminate into this class.

At last, one cannot end without asking how results extend in dimension two. Here, the main point is that this changes a lot the problem since the determinism is no more inside the tiling but only on its frontier and that it involves more complex geometric concepts since the border is no longer restricted to two points. Although very different, it may also be interesting in particular linking with the concept of self-assembly tile sets [14].

References

1. Kari, J.: The nilpotency problem of one-dimensional cellular automata. SIAM J. Comput. **21**(3), 571–586 (1992)
2. Kari, J.: Theory of cellular automata: a survey. Theor. Comput. Sci. **334**(1–3), 3–33 (2005)
3. Ganguly, N., Sikdar, B.K., Deutsch, A., Canright, G., Chaudhuri, P.P.: A survey on cellular automata. Technical report, Centre for High Performance Computing, Dresden University of Technology (2003)
4. Richard, G.: On the synchronisation problem over cellular automata. In: 34th International Symposium on Theoretical Aspects of Computer Science (2017, to appear)
5. Fischer, P.C.: Generation of primes by a one-dimensional real-time iterative array. J. ACM **12**(3), 388–394 (1965)
6. Mazoyer, J., Terrier, V.: Signals in one-dimensional cellular automata. Theor. Comput. Sci. **217**(1), 53–80 (1999)
7. Iwamoto, C., Hatsuyama, T., Morita, K., Imai, K.: Constructible functions in cellular automata and their applications to hierarchy results. Theor. Comput. Sci. **270**(1–2), 797–809 (2002)
8. Berger, R.: The undecidability of the domino problem. Ph.D. thesis, Harvard University (1964)
9. Robinson, R.: Undecidability and nonperiodicity for tilings of the plane. Inventiones Mathematicae **12**, 177–209 (1971)
10. Penrose, R.: The role of aesthetics in pure and applied mathematical research. Bull. Inst. Math. Appl. **10**(2), 266–271 (1974)
11. Culik, K., Kari, J.: An aperiodic set of Wang cubes. In: Puech, C., Reischuk, R. (eds.) STACS 1996. LNCS, vol. 1046, pp. 137–146. Springer, Heidelberg (1996). doi:10.1007/3-540-60922-9_12
12. Ben-Abraham, S.I., Gähler, F.: Covering cluster description of octagonal MnSiAl quasicrystals. Phys. Rev. B **60**, 860–864 (1999)
13. Fernique, T., Ollinger, N.: Combinatorial substitutions and sofic tilings. In: Kari, J. (ed.) Proceedings of the Second Symposium on Cellular Automata "Journeacute;es Automates Cellulaires", JAC 2010, Turku, 15–17 December 2010, pp. 100–110. Turku Center for Computer Science (2010)
14. Patitz, M.J.: An introduction to tile-based self-assembly and a survey of recent results. Nat. Comput. **13**(2), 195–224 (2014)

Some Computational Limits of Trellis Automata

Véronique Terrier[✉]

Normandie Univ, UNICAEN, ENSICAEN, CNRS, GREYC,
14000 Caen, France
veronique.terrier@unicaen.fr

Abstract. We investigate some computational limits of trellis automata. Reusing a counting argument introduced in [4], we show that:

$$\{x_1 \ldots x_n y_1 \ldots y_n : x_i y_i \in \{ab, ba, bb\} \text{ for } i = 1, \ldots, n\}$$

is not a trellis language.

1 Introduction

Trellis automata are one of the simplest parallel language recognizer. Introduced by Dyer [3], as real-time one-way bounded cellular automata, they represent a significant class of formal languages with low complexity. Notably, they are equivalent to the linear conjunctive grammars [5]. In spite of their simplicity, they have a rich computational ability and recognize various languages. In this regard, the linear context free, the visible pushdown, the poly-slender context free languages are known to be all recognized by trellis automata [1,2,6,8,11].

On the other side, some limits are known. Trellis automata are not closed under concatenation and do not contain all (and even deterministic) context-free languages. To support these claims, several languages have been shown not to be trellis languages [8–10]:

- the context free language $L_1 L_1$ square of $L_1 = \{1^k 0 u 1 0^k : k > 0, u \in \{0,1\}^*\}$
- the language $\{uvu : u, v \in \{0,1\}^*, |u| > 1\}$,
- the deterministic context free (and LL(1)) language
  $\{c^m a^{l_0} b a^{l_1} b \cdots a^{l_m} b \cdots a^{l_z} b d^n : m, n, l_i \geq 0, z \geq 1, l_m = n\}$.

The proofs rely on counting arguments which set conditions on the structure of trellis languages.

Here we will reuse another counting argument introduced in [4] in the context of functional computation, which demonstrated that the reverse operation is not realizable in minimal time on cellular automata. This argument will allow to exhibit some new prerequisite for a language to be recognized by trellis automata. As an application, we will prove that the language

© IFIP International Federation for Information Processing 2017
Published by Springer International Publishing AG 2017. All Rights Reserved
A. Dennunzio et al. (Eds.): AUTOMATA 2017, LNCS 10248, pp. 176–186, 2017.
DOI: 10.1007/978-3-319-58631-1_14

$$\{x_1 \ldots x_n y_1 \ldots y_n : x_i y_i \in \{ab, ba, bb\} \text{ for } i = 1, \ldots, n\}$$

is not a trellis language.

The paper is organized as follow. Section 2 recalls the basic definitions about trellis automata. Section 3 describes the notion of language factors diagram which can be interpreted as the language counterpart of trellis computation. Section 4 considers the patterns which may occur in the trellis computation and the ones which may occur in the factors diagrams, and also their correlation. Section 5 states a necessary condition regarding the patterns for a language to be recognizable by trellis automata. Section 6 shows that the language $\{x_1 \ldots x_n y_1 \ldots y_n : x_i y_i \in \{ab, ba, bb\} \text{ for } i = 1, \ldots, n\}$ does not fulfill such a condition.

2   Trellis Automaton

A *trellis automaton* is one of the simplest parallel language recognizer. Its underlying structure is a triangular array with sites arranged in staggered rows, as shown below.

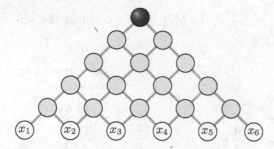

the result is read on the topmost site

the input is fed to the bottom row

A trellis automaton on an input of size 6

Formally, a trellis automaton is specified by a tuple $(Q, \Sigma, Q_{acc}, \delta)$ where

- Q is the finite set of *states*
- $\Sigma \subset Q$ is the *input* alphabet
- $Q_{acc} \subset Q$ is the set of *accepting* states
- $\delta : Q^2 \to Q$ is the *transition function*

If n is the length of w, the trellis has height n and contains on its i-th row, the $n + 1 - i$ values

$$\delta(x_1 \ldots x_i), \delta(x_2 \ldots x_{1+i}), \ldots, \delta(x_{n+1-i} \ldots x_n)$$

A trellis automaton is said to accept (resp. reject) a word $w \in \Sigma^*$, if on input w the topmost cell enters an accepting (resp. non-accepting) state.

Definition 1 (Trellis language). *A language L over an alphabet Σ is a trellis language if there exists some trellis automaton $(Q, \Sigma, Q_{acc}, \delta)$ which accepts exactly the words $w \in L$.*

Example 1. The trellis automaton $(\{a, b, c\}, \{a, b\}, \{a\}, \delta)$ accepts the set of strings of odd length whose middle symbol is a: $\text{Mi}_a = \{uav : u, v \in \{a, b\}^* \text{ and } |u| = |v|\}$

```
                    c
                  b   a
                b   c   b
              b   b   a   b
            b   b   c   b   c
          a   b   b   a   b   a
        c   b   b   c   b   c   c
      b   a   b   b   a   b   a   a
    b   c   b   b   c   b   c   c   b
  a   b   a   b   b   a   b   a   a   b
c   b   c   b   b   c   b   c   c   b   b
a   a   b   a   b   b   a   b   a   a   b   b
```

The transition function δ:

	a	b	c
a	c	b	
b	c	b	b
c		a	a

Computation on input $w = aababbabaababb$

The accepting state a marks the topmost cell of every triangle whose basis is a factor in Mi_a.

Example 2. The trellis automaton $(\{a, b, d, r\}, \{a, b\}, \{d\}, \delta)$ recognizes the set of Dyck words over $\{a, b\}$

```
                    d
                  a   b
                a   r   b
              a   r   r   b
            d   r   r   r   b
          a   b   r   r   r   b
        d   r   b   r   r   r   d
      a   b   r   b   r   r   a   b
    a   d   b   r   b   r   a   r   d
  a   a   b   b   r   b   a   r   a   b
a   d   r   d   b   r   d   r   a   d   b
a   a   b   a   b   b   a   b   a   a   b   b
```

The transition function δ:

	a	b	d	r
a	a	d	a	a
b	r	b		r
d	b		a	
r	r	b	b	r

Computation on input $w = aababbabaababb$

The accepting state d marks the Dyck words, a marks the proper prefixes of Dyck words, b marks the proper suffixes of Dyck words, r marks all the other words.

A fundamental feature of trellis automata has been noticed by Čulík:

Property 1 (Outside-context independence [1]). The computation of any word contains the computations of all its factors.

As it can be seen in Example 1 or 2, the automaton which tests the input w processes together all its factors.

3   Factors Diagram for a Language

As a matter of fact, Property 1 has strong implications on the structure of languages recognized by trellis automata. To make them explicit, let us first introduce the language counterpart of trellis computation.

Definition 2 (Factors diagram). *Let L be a language on an alphabet Σ. The indicator function of L, noted $\mathbb{1}_L$, is defined by*

$$\mathbb{1}_L : \Sigma^* \to \{0,1\}$$
$$w \to \begin{cases} 1 & if \ w \in L \\ 0 & if \ w \notin L \end{cases}$$

Let $w = x_1 \ldots x_n$ be a word. The factors diagram *of w for the language L, denoted $\Gamma_L(w)$, is a triangular array which records the values of all slices of w. If n is the length of w, the factors diagram has height n and contains on its i-th row, the $n+1-i$ values*

$$\mathbb{1}_L(x_1 \ldots x_i), \mathbb{1}_L(x_2 \ldots x_{1+i}), \ldots, \mathbb{1}_L(x_{n+1-i} \ldots x_n)$$

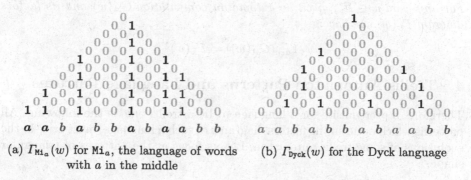

(a) $\Gamma_{\text{Mi}_a}(w)$ for Mi_a, the language of words with a in the middle

(b) $\Gamma_{\text{Dyck}}(w)$ for the Dyck language

Fig. 1. Factors diagram $\Gamma_{\mathcal{L}}(w)$ on word $w = aababbabaababb$ for the language $\mathcal{L}$

Example 3. Looking at Examples 1 and 2 where the automata evolutions on the same string are drawn, we observe that the above factors diagrams are simply projections of these automata computations. Indeed, a trellis automaton which recognizes a language L, must enter accepting states exactly on the factors belonging to L.

The following proposition formally describes the relationship between automaton evolutions and factors diagrams.

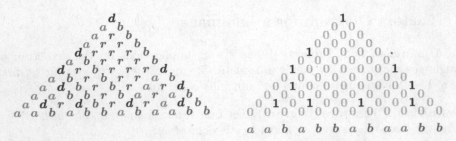

(a) The automaton computation $C_{\mathcal{A}}(w)$ (b) the factors diagram $\Gamma_{\text{Dyck}}(w)$ for the
Dyck language

Fig. 2. The factors diagram is the projection of the automaton computation

Proposition 1. *Let $\mathcal{A} = (Q, \Sigma, Q_{acc}, \delta)$ be any trellis automaton, L be the language accepted by $\mathcal{A}$ and $\mathbb{1}_{acc}$ be the indicator function of the set of accepting states:*

$$\mathbb{1}_{acc} : Q \to \{0, 1\}$$
$$q \to \begin{cases} 1 \;\; if \;\; q \in Q_{acc} \\ 0 \;\; if \;\; q \notin Q_{acc} \end{cases}$$

For any word $w \in \Sigma^*$, given its automaton computation $C_{\mathcal{A}}(w)$ and its factors diagram $\Gamma_L(w)$, we have:

$$\mathbb{1}_{acc}(C_{\mathcal{A}}(w)) = \Gamma_L(w)$$

4   Trellis Automaton Patterns and Language Patterns

Therefore, a prerequisite for a language to be a trellis one, is the following. All patterns which occur in the factors diagrams of such a language, must arise in the evolutions of some trellis automaton. Let us focus at the patterns of triangular shape.

Definition 3 (Characteristic pattern). *Let L be a language. A characteristic pattern of height h is any triangle of height h extracted from a factors diagram of L.*
$P_L(h)$ will refer to the set of all distinct characteristic patterns of height h.

Example 4. Consider Mi_a the language of strings with a in the middle. The factors diagrams of Mi_a consist of vertical stripes of only 0 or only 1.

By instance, $\begin{smallmatrix} & 1 & \\ 0 & & 0 \\ & 1 & \end{smallmatrix}$ is a characteristic pattern of height 3, but not $\begin{smallmatrix} & & 0 & \\ 0 & & & 0 \\ & 1 & \end{smallmatrix}$.

We may describe the automaton patterns in the same way:

Definition 4 (Automaton patterns). *Let $\mathcal{A} = (Q, \Sigma, Q_{acc}, \delta)$ be a trellis automaton. An automaton pattern of height h is any triangle of height h extracted from a computation of $\mathcal{A}$.*
$P_{\mathcal{A}}(h)$ will refer to the set of all distinct automaton patterns of height h.

However, trellis automata are deterministic local devices. So, for an automata pattern, the bottom row completely determines the subsequent rows. In other words, an automata pattern of height h can be viewed as a row pattern of length h complemented with its consequences.

Example 5. The automaton pattern
$$\begin{matrix} & & b & & \\ & d & & b & \\ a & & b & & b \\ d & r & & d & b \end{matrix}$$
extracted from the computation of Example 2, is entirely defined by its bottom row $\quad d \quad r \quad d \quad b \quad$ (and, of course, by the automaton rules).

As shown earlier in [9], it entails a necessary condition for a language to be recognizable by a trellis automaton, regarding to the number of its characteristic patterns:

Lemma 1. *If L is a trellis language then the number of characteristic patterns of height h, $|P_L(h)|$, is in $2^{O(h)}$.*

Proof. Assume that L is a language accepted by some trellis automaton $\mathcal{A} = (Q, \Sigma, Q_{acc}, \delta)$. According to Proposition 1, the characteristic patterns match the projection of the automaton patterns: $P_L(h) = \mathbb{1}_{acc}(P_{\mathcal{A}}(h))$. In terms of cardinal, it means that $|P_L(h)| \le |P_{\mathcal{A}}(h)|$. Moreover, the number of automaton patterns of height h is bounded by the number of distinct rows of length h where values range in Q: $|P_{\mathcal{A}}(h)| \le |Q|^h$.

Now, as the area of a characteristic pattern of height h is in $\Theta(h^2)$ and its values are 0 or 1, we can find languages whose set of characteristic patterns grows larger than $2^{O(h)}$. Using this counting argument, it has been shown that the following languages are not trellis ones:

- The context-free language $L_1 L_1$, square of the linear language $L_1 = \{1^k 0 u 1 0^k : k > 0, u \in \{0, 1\}^*\}$, since $|P_{L_1 L_1}(h)| \in 2^{\Theta(h^2)}$. See [9].
- The deterministic context-free language (and even LL(1) language) $L = \{c^m a^{l_0} b a^{l_1} b \cdots a^{l_m} b \cdots a^{l_z} b d^n : m, n, z \ge 1, l_i \ge 0, l_m = n\}$, since $|P_L(h)| \in \Omega(h!)$. See [8].

Of course, this criterion is only a necessary condition and not a sufficient one. Another drawback of this approach is that to estimate the growth rate of the characteristic patterns number of height h as h grows large, is not usually an easy task. By the way, the previous witness languages are ad hoc languages to fulfill the counting requirement. And the status of more common languages remains as yet unknown. Two candidates are currently mentioned:

- The balanced language over $\{a, b\}$ defined as the set of strings with the same number of symbols a and b:

$$\mathbf{Eq} = \{w \in \{a, b\}^* : \sharp_a(w) = \sharp_b(w)\}$$

– The copy language defined as the set of words repeated twice:

$$\text{Copy} = \{ww : w \in \{a,b\}^*\}$$

Here we will look at the language Mi_aMi_a and the variant Mi_aMi_b where Mi_a (resp. Mi_b) stands for the set of odd length words with a (resp. b) in the middle:

$$\text{Mi}_a = \{xay \in \{a,b\}^* : |x| = |y|\}$$

Making use of an approach introduced in [4], we will show that Mi_aMi_a and Mi_aMi_b are not trellis languages. But although they are closely related to the Copy language and its negative variant:

$$\text{Copy} = (\text{Mi}_a\text{Mi}_b)^{\complement} \cap (\text{Mi}_b\text{Mi}_a)^{\complement} \cap \{aa, ab, ba, bb\}^*$$

$$\{w\overline{w} : w \in \{a,b\}^*\} = (\text{Mi}_a\text{Mi}_a)^{\complement} \cap (\text{Mi}_b\text{Mi}_a)^{\complement} \cap \{aa, ab, ba, bb\}^* = \text{Eq} \cap (\text{Mi}_a\text{Mi}_a)^{\complement}$$

it will not allow us to determine whether they are trellis languages or not.

5   Counting Argument

Here we will focus on a subfamily of the characteristic patterns composed of horizontal stripes.

Definition 5 (Stripes patterns). *A stripes pattern is a characteristic pattern such that all the values within each row are equal. The characteristic string of a stripes pattern of height h is the binary string $c = c_1 \cdots c_h$ of length h where c_i is the 0 or 1 value of the i-th row of the stripes pattern.*

An automaton pattern π would be said to have a characteristic string c if its projection $\mathbb{1}_{acc}(\pi)$ is a stripes pattern of characteristic c.

Note that the characteristic string completely characterizes the stripes pattern. And so, whatever the language, the number of its stripes patterns of height h is bounded by 2^h. Regardless of the fact that the subfamily of stripes patterns is not so large and even within the bound defined in Lemma 1, it has been proved that any trellis automaton could not display all of them:

Proposition 2 (Grandjean, Richard, Terrier [4]). *For any trellis automaton $\mathcal{A}$, there exist some stripes patterns which never occur in the space-time diagrams of $\mathcal{A}$.*

Along the same lines, Proposition 2 could be refined to deal with languages exhibiting not necessarily all stripes patterns.

Definition 6. *For any language L, $\mathcal{C}_L$ will refer to the set of characteristic strings whose corresponding stripes patterns occur in L.*
Give, any subset $\mathcal{F} \subset \mathcal{C}_L$, the integer $\alpha_h^{\mathcal{F}}$ will refer to the minimal number of double length extensions of every string of length 2^h within $\mathcal{F}$:

$$\alpha_h^{\mathcal{F}} = \min_{c \in \mathcal{F}, |c|=2^h} (|\{d \in \mathcal{F} : c \text{ is a prefix of } d \text{ and } |d| = 2|c|\}|)$$

Proposition 3.   *If L is a language which admits a subset $\mathcal{F}$ of characteristic strings such that the sequence $(\alpha_h^{\mathcal{F}})$ is monotonic and divergent, then L is not a trellis language.*

The counting argument used to prove the proposition is based on the next technical fact.

Fact 1. *Let (α_h) be any monotonic sequence of positive integers which is divergent: $\alpha_{h+1} \geq \alpha_h$ for all h, and $\lim_{h\to\infty} \alpha_h \to \infty$. Let C be any positive constant. Then the sequence (u_h) defined recursively by:*

$$u_0 = C \quad and \quad u_{h+1} = \frac{u_h^2}{\alpha_h}$$

converges to 0.

Proof. First, observe that

$$u_h = C^{2^h} / \prod_{i=0}^{h-1} \alpha_i^{2^{h-i-1}}$$

Second, by assumption, there exists an index H such that $\alpha_h \geq C + 1$, for all $h \geq H$. Then for $h \geq H$,

$$u_h \leq C^{2^h} / \prod_{i=H}^{h-1} (C+1)^{2^{h-i-1}} = C^{2^h} / (C+1)^{2^{h-H}}$$

So the sequence (u_h) converges to 0.

Proof (Proposition 3). Assume that L is a language accepted by some trellis automaton $\mathcal{A} = (Q, \Sigma, Q_{acc}, \delta)$. We will construct a sequence of strings w_i of length 2^i belonging to $\mathcal{F}$ such that the number of automaton patterns with characteristic w_i is bounded by u_i. Then, according to Fact 1, we will have $u_I < 1$ for I large enough. That means there will be no automaton pattern with characteristic u_I and hence w_I would not be a characteristic string of L. Thus the assumption that L is a trellis language, would lead to a contradiction.

The construction of the sequence of strings w_i is done by recurrence:

The base case. For $i = 0$, the automaton patterns of height 1 are reduced to one site and their number is bounded by the cardinal of Q. So there are at most $C = |Q|$ automaton patterns with characteristic string 0 or 1. Let set w_0 be a string of length 1 belonging to $\mathcal{F}$ and u_0 be $|Q|$.

The inductive step. Consider all automaton patterns of height 2^{i+1} having a characteristic string within $\mathcal{F}$ which is a double length extension of w_i. As depicted in Fig. 3, we can divide such a kind of pattern in four sub-patterns X, Y, Z and T where X, Y and Z are of height 2^i and also where X and Y share the characteristic w_i. By recurrence assumption, the number of automaton patterns of characteristic w_i is bounded by u_i and so the number of couples (X, Y) is at most u_i^2. Furthermore the sub-patterns Z and T depend only on X and Y. That is to say the number of automaton patterns whose characteristic strings are extensions of w_i is bounded by u_i^2. Now, since the minimal number of extensions of w_i within $\mathcal{F}$ is α_i, the average number of automaton patterns per extension is bounded by $u_{i+1} = u_i^2 / \alpha_i$. In other words, there is one extension w_{i+1} of w_i with length 2^{i+1} and belonging to $\mathcal{F}$ such that the number of automaton patterns with characteristic w_{i+1} is bounded by u_{i+1}.

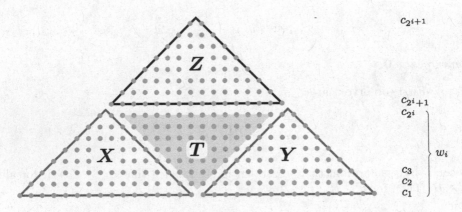

Fig. 3. Subdivision of an automaton pattern in four patterns X, Y, Z and T

6 Some Non Trellis Language

Now we will apply the previous criterion to show that the language $\text{NO}_{aa} = \{x_1 \ldots x_n y_1 \ldots y_n : x_i y_i \in \{ab, ba, bb\} \text{ for } i = 1, \ldots, n\} \cup \{w \in \{a, b\}^* : w \text{ is of odd length}\}$ is not a trellis language. As an aside, notice that NO_{aa} is not a context-free language although its complement $\text{Mi}_a \text{Mi}_a$ is a context-free one.

As preliminary, let us look at an example. Figure 4 depicts the factors diagram on the input word $^\omega bab^{12} abbbabaaabbab^\omega$. The dark sites mark the 0 values (i.e., the factors not in NO_{aa}), the light sites mark the 1 values. We observe that the black horizontal stripes in the upper part match the symbols a of the input.

More generally, the NO_{aa} factors diagrams exhibit the following stripes patterns:

Fact 2. *For any binary string $c_1 c_2 \ldots c_k$, there exists a stripes pattern of NO_{aa} with characteristic string $c_1 1 c_2 1 \ldots 1 c_k 1$.*

Proof. Given any binary string $c_1 c_2 \ldots c_k$ of length k, we consider the word $w = b^{m+k-1} ab^m x_1 \cdots x_k b^{m+k-1}$ where m is any integer greater than k and, the

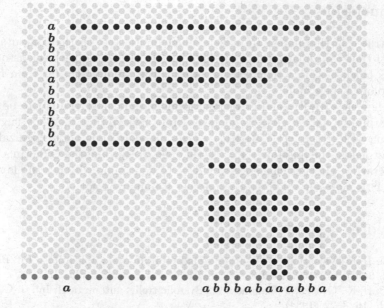

Fig. 4. The factors diagram on $^\omega bab^{12}abbbabaaabbab^\omega$ for the language NO_{aa}

symbols x_i are a if $c_i = 0$ and b otherwise. As it is defined, each symbol x_i decides whether the $m + i$ factors with length $2(m + i)$ of $b^{m+i-1}ab^m x_1 \cdots x_k b^{m+2i-k-1}$ are all in NO_{aa} (in case of $x_i = b$) or are all outside of NO_{aa} (in case of $x_i = a$). Therefore the factors diagram of w contains on its $2(m + i)$-row a sequence of $m + i$ consecutive values c_i and that for all $i = 1, \cdots, k$. Besides, all values of the odd rows are 1 since any odd length factor is in NO_{aa}. At last, choosing m large enough, we can extract from the factors diagram of w a stripes pattern of characteristic $c_1 1 c_2 1 \ldots 1 c_k 1$.

Proposition 4. *The language* $\mathrm{NO}_{aa} = \{x_1 \ldots x_n y_1 \ldots y_n : x_i y_i \in \{ab, ba, bb\}$ *for* $i = 1, \ldots, n\} \cup \{w \in \{a, b\}^* : w$ *is of odd length*$\}$ *is not a trellis language.*

Proof. According to Fact 2, every string of $\mathcal{F} = \{01, 11\}^*$ is a characteristic string of NO_{aa}. Besides, within $\mathcal{F}$, every string $c_1 1 \ldots c_{2^h-1} 1$ of length 2^h is the prefix of 2^{h-1} strings of double length: $\{c_1 1 \ldots c_{2^h-1} 1 e_1 1 \ldots e_{2^h-1} 1 : e_1, \ldots, e_{2^h-1} \in \{0, 1\}\} \subset \mathcal{F}$. Hence $\alpha_h^{\mathcal{F}} = 2^{h-1}$ and so the sequence $(\alpha_h^{\mathcal{F}})$ is monotonic and divergent. Then it follows from Proposition 3 that NO_{aa} is not a trellis language.

As a matter of fact, it can be shown in the same way that the language $\mathrm{NO}_{ab} = \{x_1 \ldots x_n y_1 \ldots y_n : x_i y_i \in \{aa, ba, bb\}$ for $i = 1, \ldots, n\} \cup \{w \in \{a, b\}^* : w$ is of odd length$\}$ is not a trellis language. At the same time, neither $\mathrm{Mi}_a \mathrm{Mi}_a$ nor $\mathrm{Mi}_a \mathrm{Mi}_b$ are trellis languages.

7   Conclusion

As illustrated in this paper, to make explicit limitations on the computational ability of trellis automata, the analysis of the characteristic patterns associated to trellis languages, is a significant approach. But we are still far from having fully exploited such tools.

The language Mi_aMi_a and its derived forms have been shown not to be trellis ones. Despite the fact it gives us good reason to believe that the Copy language, coinciding with $(\text{Mi}_a\text{Mi}_b)^{\complement} \cap (\text{Mi}_b\text{Mi}_a)^{\complement} \cap \{aa, ab, ba, bb\}^*$, is not recognizable by trellis automata, the question remains still open. Regarding the Okhotin's grammars hierarchy, another challenge would be to determine whether the language $(\text{Mi}_a\text{Mi}_a)^{\complement}$ is representable by a conjunctive grammar or not [7].

References

1. Čulík II, K.: Variations of the firing squad problem and applications. Inf. Process. Lett. **30**(3), 152–157 (1989)
2. Čulík II, K., Gruska, J., Salomaa, A.: Systolic trellis automata II. Int. J. Comput. Math. **16**, 3–22 (1984)
3. Dyer, C.R.: One-way bounded cellular automata. Inf. Control **44**(3), 261–281 (1980)
4. Grandjean, A., Richard, G., Terrier, V.: Linear functional classes over cellular automata. In: Formenti, E. (ed.), Proceedings AUTOMATA & JAC 2012, pp. 177–193 (2012)
5. Okhotin, A.: Automaton Representation of Linear Conjunctive Languages. In: Ito, M., Toyama, M. (eds.) DLT 2002. LNCS, vol. 2450, pp. 393–404. Springer, Heidelberg (2003). doi:10.1007/3-540-45005-X_35
6. Okhotin, A.: On the equivalence of linear conjunctive grammars and trellis automata. RAIRO Informatique Théorique et Applications **38**(1), 69–88 (2004)
7. Okhotin, A.: Conjunctive and boolean grammars: the true general case of the context-free grammars. Comput. Sci. Rev. **9**, 27–59 (2013)
8. Okhotin, A.: Input-driven languages are linear conjunctive. Theoret. Comput. Sci. **618**, 52–71 (2016)
9. Terrier, V.: On real time one-way cellular array. Theoret. Comput. Sci. **141**(1–2), 331–335 (1995)
10. Terrier, V.: Language not recognizable in real time by one-way cellular automata. Theoret. Comput. Sci. **156**(1–2), 281–287 (1996)
11. Terrier, V.: Recognition of poly-slender context-free languages by trellis automata. Theoret. Comput. Sci. (2017)

Turing-Completeness of Asynchronous Non-camouflage Cellular Automata

Tatsuya Yamashita[1(✉)], Teijiro Isokawa[2], Ferdinand Peper[3],
Ibuki Kawamata[4], and Masami Hagiya[1]

[1] University of Tokyo, Tokyo, Japan
`t.yamashita@is.s.u-tokyo.ac.jp`
[2] University of Hyogo, Kobe, Japan
[3] NICT and Osaka University, Osaka, Japan
[4] Tohoku University, Sendai, Japan

Abstract. Asynchronous Boolean totalistic cellular automata have recently attracted attention as promising models for the implementation of reaction-diffusion systems. It is unknown, however, to what extent they are able to conduct computation. In this paper, we introduce the so-called *non-camouflage property*, which means that a cell's update is insensitive to neighboring states that equal its own state. This property is stronger than the *Boolean totalistic property*, which signifies the existence of states in a cell's neighborhood, but is not concerned with how many cells are in those states. We argue that the non-camouflage property is extremely useful for the implementation of reaction-diffusion systems, and we construct an asynchronous cellular automaton with this property that is Turing-complete. This indicates the feasibility of computation by reaction-diffusion systems.

1 Introduction

Recent efforts towards the molecular implementation of reaction-diffusion systems have resulted in the characterization of cellular automata that are suitable for this purpose [2,3]. A possible implementation for this kind of CA uses a porous material, such as an *alginate* or *polyacrylamide gel*, as the framework of the cellular space. In this type of material, many small (millimeter scale) holes are arranged as a lattice, each of which is employed as a cell, with boundaries made of this material. Artificial DNA molecules are then used to represent cell states, whereby their chemical reactions represent transition rules acting upon these states. These DNA molecules are broadly divided into two types according to their size. Small molecules are able to pass through the porous material at a cell's boundary, but big molecules are not. Big molecules are thus suitable to be used for representing the state of a cell, whereas small molecules can act as transmitters to neighboring cells. The reactions between molecules are designed according to the transition rule of the implemented CA. A computation on the CA is then initiated by injecting the designed molecules into each cell (hole)

A. Dennunzio et al. (Eds.): AUTOMATA 2017, LNCS 10248, pp. 187–199, 2017.
DOI: 10.1007/978-3-319-58631-1_15

depending on the initial state of the CA. Computational cellular systems created by the above procedure are called *Gellular Automata* [1,6].

In the scheme outlined above, the implemented CA must satisfy certain requirements to allow it to exploit the characteristics of molecular implementations. Since it is difficult to synchronize the chemical reactions in all cells, the CA should be asynchronous, rather than an ordinary synchronous CA. In addition, it is also difficult for reaction-diffusion systems to recognize the direction from which DNA molecules have come, so, rather than identifying the state of each neighboring cell, we merely use the number of neighboring cells in certain states (totalistic CA). This is not sufficient, though, since it is quite difficult to estimate the amount of diffused DNA molecules in cells, and even to establish how many neighboring cells are in a certain state. For this reason, it was proposed to refine the totalistic CA to so-called *Boolean totalistic CA* [2], in which the mere presence and absence of states among the neighbors of a cell are sufficient in the definition of transition rules.

There is an additional difficulty in this scheme, however. Imagine that a cell in a certain state is supposed to change to another state if there is a neighboring cell whose state is identical to the current state of the cell. Such a transition rule is allowed in a conventional asynchronous Boolean totalistic CA. However, in a reaction-diffusion implementation, a cell cannot recognize the existence of a neighboring cell in the same state since the cell itself is emitting the transmitter indicating its state. To resolve this difficulty, we define the *non-camouflage property* in this paper, which in effect ignores a state of a cell's neighbor if the state equals the state of the cell itself. This property is stronger than the Boolean totalistic property. We present an asynchronous non-camouflage CA and prove that it is Turing-complete.

This paper is organized as follows. Section 2 gives the formal definitions of the used concepts. This is followed by a description of the proposed CA in Sect. 3, and the proof that it is Turing-universal. This paper finishes with a discussion in Sect. 4.

2 Preliminaries

2.1 Asynchronous CA

In this paper, we follow the terminology used in [5]. State transition systems, which are pairs of a set and a binary relation on the set, are called *state-systems*. We then define synchronous and asynchronous CA as state-systems. Note that ordinary CA are synchronous CA while in this paper, we only deal with asynchronous CA.

Definition 1 (State-system). *A state-system A is a pair $A = (T, \rightarrow)$, where T is a set of states, and $\rightarrow \in T \times T$ is a binary relation meaning state transition.*

For $(t_1, t_2) \in \rightarrow$, we write $t_1 \rightarrow t_2$ and say "the state t_1 is changed to t_2." Let $t_0, t_n \in T$ be states. If there are states $t_1, \ldots, t_{n-1}$ and $t_i \rightarrow t_{i+1}$ holds for each $i = 0, \ldots, n-1$, we write $t_0 \rightarrow^* t_n$.

To prove that a state-system B is computationally more powerful than a state-system A or equally powerful as A, we need to show that B can simulate A. Here is the definition of simulation derived from [5] but slightly modified for our purpose.

Definition 2 (Simulation). *A state-system $B = (T_B, \to_B)$ simulates a state-system $A = (T_A, \to_A)$ if there is a function $F : T_A \to T_B$ and*

(i) $\forall t_1, t_2 \in T_A.\ t_1 \to_A t_2 \implies \forall t' \in T_B.\ F(t_1) \rightsquigarrow_F t' \implies t' \rightsquigarrow_F F(t_2)$
(ii) $\forall t_1, t_2 \in T_A.\ F(t_1) \to_B^* F(t_2) \implies t_1 \to_A^* t_2$

where $\rightsquigarrow_F$ denotes the binary relation over T_B that is defined as

$$t' \rightsquigarrow_F t'' \iff \exists n \in \mathbb{N}.\ \exists t'_0, \dots, t'_n \in T_B.$$
$$t' = t'_0 \to_B t'_1 \to_B \cdots \to_B t'_n = t'' \land \forall i \in [1, n-1].\ t'_i \notin F(A).$$

The function F in this definition is called a *simulation function* (or simply a *simulation*) of A by B.

Intuitively, (i) means that for any transition $t_1 \to_A t_2$, there is a sequence of transitions from $F(t_1)$ to $F(t_2)$, which does not go through the image of A by F. Since the binary relation $\rightsquigarrow_F$ is reflexive, (i) implies

$$\forall t_1, t_2 \in T_A.\ t_1 \to_A t_2 \implies F(t_1) \to_B^* F(t_2).$$

This corresponds to one of the original conditions of simulation in [5]. (ii) means that any sequence of transitions from $F(t_1)$ to $F(t_2)$ in B corresponds to a sequence of transitions from t_1 to t_2 in A.

Next, we define a CA. A CA is a state-system whose states can be considered as an arrangement in a certain topology[1] of cell states, whereby the transition relation is determined by applying a local transition rule to each cell simultaneously. We define asynchronous CA, which are discussed mainly in this paper.

Definition 3 (Asynchronous CA). *A state-system $A = (T, \to)$ is called an asynchronous CA if the following two conditions are satisfied.*

(i) There is a set S_A of cell states such that $T = S_A^{\mathbb{Z} \times \mathbb{Z}}$.
(ii) There is a function $f_A : S_A^5 \to S_A$ such that for any two states $t_1, t_2 \in T$,

$$t_1 \to t_2 \iff \forall x, y \in \mathbb{Z}.\ (t_2(x, y) = f_A(t_1(x, y), t_1(x+1, y), t_1(x, y+1),$$
$$t_1(x-1, y), t_1(x, y-1))$$
$$\lor\ t_2(x, y) = t_1(x, y)).$$

For an asynchronous CA A, S_A and f_A are called the *space of cell states* and the *transition rule*, respectively.

---

[1] We discuss CA with the two-dimensional lattice arrangement using the von Neumann neighborhood.

2.2   Requirements

Here, we define the requirements for asynchronous CA to be implemented by reaction-diffusion systems. The following definitions are about asynchronous CA but the word "outer totalistic" can also be applied to synchronous CA.

Definition 4 (Outer totalistic). *The function* $tot : S_A^5 \to S_A \times \mathbb{N}^{S_A}$ *is defined by* $tot(s_0, s_1, s_2, s_3, s_4) = (s_0, h)$, *where*

$$h(s) = \sum_{k=1}^{4} g(s_k, s), \ g(s, s') = \begin{cases} 1 \ if \ s = s' \\ 0 \ otherwise \end{cases}$$

are functions $h : S_A \to \mathbb{N}$ *and* $g : S_A \times S_A \to \mathbb{N}$. *An asynchronous CA A is (outer) totalistic if there is a function* $f'_A : S_A \times \mathbb{N}^{S_A} \to S_A$ *and* $f_A = f'_A \circ tot$.

Definition 5 (Boolean totalistic). *Let the symbol* 2 *denote the set of Boolean values. The function* $bol : S_A^5 \to S_A \times 2^{S_A}$ *is defined by* $bol(s_0, s_1, s_2, s_3, s_4) = (s_0, h)$, *where*

$$h(s) = \bigvee_{k=1}^{4} (s_k = s)$$

is a function $h : S_A \to 2$. *An asynchronous CA A is Boolean totalistic if there is a function* $f''_A : S_A \times 2^{S_A} \to S_A$ *and* $f_A = f''_A \circ bol$.

If an asynchronous CA A is Boolean totalistic, the transition rule is determined by the function $f''_A : S_A \times 2^{S_A} \to S_A$. We identify the transition rule f_A with f''_A and represent it by a list of the form like $s_0(s_i, \ldots, \neg s_j, \ldots) \to s'_0$. This expression means that a cell with state s_0 can be changed to state s'_0 when in its neighborhood there are cells with state $s_i, \ldots$ and there is no cell with state $s_j, \ldots$. If there is no form whose left-hand side is applicable to a situation $(s_0, \ldots, s_4)$, then f''_A returns s_0. A cell in such a situation $(s_0, \ldots, s_4)$ will not change its state by a transition of A. If there is more than one form whose left-hand side are applicable to a situation $(s_0, \ldots, s_4)$, then f''_A follows the first one.

Since there is a function $g : S_A \times \mathbb{N}^{S_A} \to S_A \times 2^{S_A}$ such that $bol = g \circ tot$, an asynchronous Boolean totalistic CA is outer totalistic. In other words, the Boolean totalistic property is stronger than the outer totalistic property.

As discussed in Sect. 1, the Boolean totalistic property is still not strong enough for our purposes. We define stronger property, non-camouflage.

Definition 6 (Non-camouflage). *An asynchronous Boolean totalistic CA A is non-camouflage if its transition rule* f''_A *satisfies*

$$\forall s_0 \in S_A. \ \forall h_0, h_1 \in 2^{S_A}.$$
$$(\forall s \in S_A. \ s \neq s_0 \implies h_0(s) = h_1(s)) \implies f''_A(s_0, h_0) = f''_A(s_0, h_1),$$

If an asynchronous Boolean totalistic CA is non-camouflage, it is called an asynchronous non-camouflage CA.

2.3   Priese System

In Sect. 3, we construct an asynchronous non-camouflage CA, and prove that it is Turing-complete. In [5], Priese defined a Turing-complete system, which we call a *Priese System* in this paper and define in this subsection. Note that a Priese System is suitable for our purposes because it does not require synchronization of its elements.

First, we define *s-automata* (named after sequential automata).

Definition 7 (s-automaton). *A tuple $A = (I, O, S, \rightarrow)$ is called an s-automaton if*

(i) I and O are finite sets with $I \cap O = \emptyset$.
(ii) S is a set.
(iii) $\rightarrow \subset (I \times S) \times (O \times S)$ is a transition relation.

An s-automaton $A = (I, O, S, \rightarrow)$ is thus a machine that has a set I of input terminals, a set O of output terminals, and a set S of inner states. Let $x \in I$, $y \in O$ and $s, s' \in S$ be an input terminal, an output terminal and inner states, respectively. $((x, s), (y, s')) \in \rightarrow$ is denoted as $(x, s) \rightarrow (y, s')$. This state transition is interpreted as follows. If the input terminal x receives a signal and the s-automaton A is in the inner state s, then A can remove the signal on the input terminal x, change its inner state from s to s' and add a signal to the output terminal y.

An s-automaton $A = (I, O, S, \rightarrow)$ can be considered as a state-system $((I \sqcup O) \times S, \rightarrow)^2$.

Two s-automata called K and E are used to define the Priese System. The s-automaton $K = (I_K, O_K, S_K, \rightarrow_K)$ is defined by

$$I_K = \{0, 1\}, O_K = \{2\}, S_K = \{0\}, \rightarrow_K = \{((0, 0), (2, 0)), ((1, 0), (2, 0))\}.$$

This s-automaton has two input terminals. Whichever input terminal receives a signal, it flows to the unique output terminal. The s-automaton $E = (I_E, O_E, S_E, \rightarrow_E)$ is defined by

$$I_E = \{s, t\}, O_E = \{s', t^u, t^d\}, S_E = \{u, d\},$$
$$\rightarrow_E = \{((s, u), (s', d)), ((s, d), (s', u)), ((t, u), (t^u, u)), ((t, d), (t^d, d))\}.$$

This s-automaton has two inner states. When a signal arrives at the input terminal t, it flows to the output terminal t^u or t^d depending on the inner state of the s-automaton E. When a signal arrives at the input terminal s, it flows to the output terminal s' and the inner state is flipped at the same time.

Both of the s-automata K and E are too simple to simulate a universal Turing machine on their own, but a system constructed by connecting them turns out to be powerful enough. To connect s-automata each other, we define two operations over s-automata, product and feed-back.

The *product* of s-automata A and B is the s-automaton given by arranging them in a parallel configuration.

---

[2] $\sqcup$ denotes a disjoint union.

Definition 8 (Product). *Let s-automata* $A = (I_A, O_A, S_A, \rightarrow_A)$, $B = (I_B, O_B, S_B, \rightarrow_B)$ *be given. The product* $A \otimes B$ *is the s-automaton* $(I_A \sqcup I_B, O_A \sqcup O_B, S_A \times S_B, \rightarrow_{A \otimes B})$ *with* $\rightarrow_{A \otimes B} = \{((x, (s, t)), (y, (s', t))) \mid (x, s) \rightarrow_A (y, s'), t \in S_B\} \cup \{((x, (s, t)), (y, (s, t'))) \mid (x, t) \rightarrow_B (y, t'), s \in S_A\}$.

Let s-automaton A be given. The *feed-back* of the output terminal y to the input terminal x is the s-automaton given by connecting y to x.

Definition 9 (Feed-back). *Let an s-automaton* $A = (I_A, O_A, S_A, \rightarrow_A)$, *an input terminal* $x \in I_A$, *and an output terminal* $y \in O_A$ *be given. The feedback* A_y^x *of the output terminal* y *to the input terminal* x *is the s-automaton* $(I_A \backslash \{x\}, O_A \backslash \{y\}, S_A, \rightarrow_{A_y^x})$ *with*

$$\rightarrow_{A_y^x} = Cl(\rightarrow_A \cup \{((y, s), (x, s)) \mid s \in S_A\}) \cap ((I_A \backslash \{x\} \times S_A) \times (O_A \backslash \{y\} \times S_A)),$$

where Cl denotes the transitive and reflexive closure of a binary relation.

If one wants to make a machine that is made of s-automata $A_1, \ldots, A_n$, he or she can put them in parallel by the operation product and connect them to each other by the feed-back operation. The class of s-automata generated by such operations is called *Normed Networks.*

Definition 10 (Normed Network). *Let s-automata* $A_1, \ldots, A_n$ *be given. The Normed Network over* $A_1, \ldots, A_n$ *is the smallest set of s-automata that*

(i) contains $A_1, \ldots, A_n$ *and*
(ii) is closed under feed-back and product.

Priese System is defined as the Normed Network over s-automata K and E. It is known that any finite-state s-automaton belongs to Priese System [5]. Next we define the infinite chain made of two s-automata A and B, where A and infinite copies of B are connected in sequence.

Definition 11 (Infinite chain). *Let s-automata* $A = (I_A \sqcup \bar{I}_A, O_A \sqcup \bar{O}_A, S_A, \rightarrow_A)$ *and* $B = (I_B \sqcup \bar{I}_B, O_B \sqcup \bar{O}_B, S_B, \rightarrow_B)$ *be given and* $|\bar{I}_A| = |O_B| = |\bar{I}_B| = m$, $|\bar{O}_A| = |\bar{O}_B| = |I_B| = n$. *Let* $B^{(i)} = (I_B^{(i)} \sqcup \bar{I}_B^{(i)}, O_B^{(i)} \sqcup \bar{O}_B^{(i)}, S_B^{(i)}, \rightarrow_B^{(i)})$ *be disjoint copies of* B *and* $\bar{I}_A = \{\bar{x}_1, \ldots, \bar{x}_m\}$, $\bar{O}_A = \{\bar{y}_1, \ldots, \bar{y}_n\}$, $I_B^{(i)} = \{x_1^{(i)}, \ldots, x_n^{(i)}\}$, $O_B^{(i)} = \{y_1^{(i)}, \ldots, y_m^{(i)}\}$, $\bar{I}_B^{(i)} = \{\bar{x}_1^{(i)}, \ldots, \bar{x}_m^{(i)}\}$, $\bar{O}_B^{(i)} = \{\bar{y}_1^{(i)}, \ldots, \bar{y}_n^{(i)}\}$. *The infinite chain made of* A *and* B *is an s-automaton* $(S, I, O, \rightarrow)$ *where* $S = \{(s, t_0, t_1, \ldots) \mid s \in S_A, t_i \in S_B^{(i)}\}$, $I = I_A$, $O = O_A$, *and*

$$\rightarrow = Cl(\{((x, (s, t_0, t_1, \ldots)), (y, (s', t_0, t_1, \ldots))) \mid (x, s) \rightarrow_A (y, s')\}$$
$$\cup \{((x, (s, t_0, \ldots, t_i, \ldots)), (y, (s, t_0, \ldots, t_i', \ldots))) \mid (x, t_i) \rightarrow_B^{(i)} (y, t_i')\}$$
$$\cup \{((\bar{y}_k, u), (x_k^{(0)}, u)) \mid \bar{y}_k \in \bar{O}_A, x_k^{(0)} \in I_B^{(0)}\}$$
$$\cup \{((y_k^{(0)}, u), (\bar{x}_k, u)) \mid y_k^{(0)} \in O_B^{(0)}, \bar{x}_k \in \bar{I}_A\}$$
$$\cup \{((\bar{y}_k^{(i)}, u), (x_k^{(i+1)}, u)) \mid \bar{y}_k^{(i)} \in \bar{O}_B^{(i)}, x_k^{(i+1)} \in I_B^{(i+1)}\}$$
$$\cup \{((y_k^{(i+1)}, u), (\bar{x}_k^{(i)}, u)) \mid y_k^{(i+1)} \in O_B^{(i+1)}, \bar{x}_k^{(i)} \in \bar{I}_B^{(i)}\})$$
$$\cap ((I \times S) \times (O \times S)).$$

It is also known that any computation of a Turing machine from a finite initial configuration can be simulated by a computation of an infinite chain of finite-state s-automata. An argument showing this fact can be found, for instance, in [4].

3 Proposing Asynchronous Non-camouflage CA

3.1 Proposed CA M

In this subsection, we present an asynchronous non-camouflage CA.

Table 1 shows the transition rule of our asynchronous non-camouflage CA in the asynchronous Boolean totalistic form. For simplicity, we call this asynchronous CA M. By the definition of asynchronous CA, M is a state system $M = (S_M^{\mathbb{Z} \times \mathbb{Z}}, \to_M)$. The number of cell states becomes $|S_M| = 21$ by adding a state 0 to the 20 states appearing in the table of transition rules.

Table 1. The transition rule of M.

1	$1\,(2, \neg C_1, \neg E_0) \to Z$	13	$Z\,(3, U_0, \neg L, \neg D_0, \neg E_1) \to u$
2	$1\,(W, \neg K) \to Z$	14	$Z\,(3, D_0, \neg L, \neg U_0, \neg E_1) \to d$
3	$1\,(u, \neg E_0) \to Z$	15	$C_0\,(Z) \to C_1$
4	$1\,(d, E_0) \to Z$	16	$C_1\,(4) \to C_0$
5	$2\,(3, Z, \neg U_0, \neg D_0) \to Y$	17	$W\,(3, Z) \to Y$
6	$3\,(4, Y) \to X$	18	$U_0\,(1, 2, E_0) \to D_1$
7	$4\,(X, \neg C_1, \neg U_1, \neg D_1) \to 1$	19	$D_1\,(1, 4, E_0) \to D_0$
8	$X\,(1, Y, \neg 4) \to 4$	20	$D_0\,(1, 2, E_0) \to U_1$
9	$Y\,(4, Z) \to 3$	21	$U_1\,(1, 4, E_0) \to U_0$
10	$Z\,(3, \neg C_0, \neg L, \neg D_0, \neg U_0) \to 2$	22	$u\,(3, Z) \to Y$
11	$Z\,(3, L) \to W$	23	$d\,(3, Z) \to Y$
12	$Z\,(3, E_1, \neg L) \to 2$		

In the table of transition rules, state s_0 of each rule does not appear in the bracket $(s_i, \ldots, \neg s_j, \ldots)$. This fact implies that M is non-camouflage.

Since a cell in the state 0 will never be changed to another state or influence transitions of neighboring cells, we assume that almost all cells are in the cell state 0 and such cells are not drawn in the figures.

3.2 Simulation of s-automaton

In this subsection, we show that M simulates any s-automaton belonging to the Priese System. M can also simulate an infinite chain made of two s-automata in Priese System with an initial configuration that is periodic except for a finite area. The Turing-completeness of M follows from this fact.

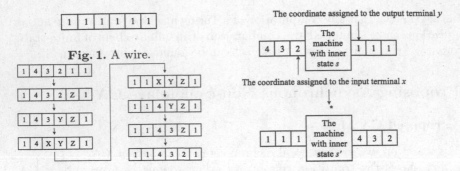

Fig. 1. A wire.

Fig. 2. A signal on a wire.   **Fig. 3.** An image of wire-based simulation.

Wires and Signals. Cells with the state $1 \in S_M$ extending linearly are called a wire. Figure 1 shows an image of a wire. Three cells in the states $2, 3, 4 \in S_M$ arranged in this order are called a signal. This order makes a signal directed. A signal on a wire progresses along the wire (see Fig. 2). A signal on a finite wire reaches the end of the wire in a finite number of transitions if there is no influence from the outside on the wire.

Let $A = (I_A, O_A, S_A, \to_A)$ be an s-automaton. Recall that A is simulated as a state-system $((I_A \sqcup O_A) \times S_A, \to_A)$. A state $(io, s) \in (I_A \sqcup O_A) \times S_A$ of A is regarded as a situation in which a machine with an inner state s has a signal on a terminal io. Wire-based simulation functions are simulation functions of s-automata by M, and represent such situations. Figure 3 shows how a wire-based simulation represents a transition $(x, s) \to (y, s')$ of an s-automaton with one input terminal x and one output terminal y.

Definition 12 (Wire-based simulation). *Let $A = (I_A, O_A, S_A, \to_A)$ be an s-automaton and $F : (I_A \sqcup O_A) \times S_A \to \wp(S_M^{\mathbb{Z} \times \mathbb{Z}})$ be a simulation function of A by M. F is called a wire-based simulation function if there are functions $G : S_A \to S_M^{\mathbb{Z} \times \mathbb{Z}}$, $g : I_A \sqcup O_A \to \mathbb{Z} \times \mathbb{Z}$ and integers $x_{min}, x_{max}, y_{min}, y_{max} \in \mathbb{Z}$ such that*

(i) for each state $s \in S_A$ of A, there is no transition from $G(s)$,
(ii) $\forall s \in S_A. \forall x, y \in \mathbb{Z}. G(s)(x, y) = 0 \vee (x_{min} \leq x \leq x_{max} \wedge y_{min} \leq y \leq y_{max})$,
(iii) $\forall io \in I_A \sqcup O_A. \forall x, y \in \mathbb{Z}. g(io) = (x, y) \implies$
  $((x = x_{min} - 1 \vee x = x_{max} + 1) \wedge (y_{min} < y < y_{max}))$
  $\vee((y = y_{min} - 1 \vee y = y_{max} + 1) \wedge (x_{min} < x < x_{max}))$,
(iv) for each input terminal $i \in I_A$ and state $s \in S_A$ of A, the state $F(i, s)$ is identical with the state constructed by replacing a series of three cells with states $(0, 0, 0)$ in $G(s)$, which is extended toward the outside from the coordinate $g(i)$, with an inwardly directed signal $(2, 3, 4)$, and replacing series of three cells with states $(0, 0, 0)$, which are extended toward the outside from the coordinate $g(io)$ for each terminal $io \in (I_A \setminus \{i\}) \sqcup O_A$, with wires $(1, 1, 1)$, and

(v) for each output terminal $o \in O_A$ and state $s \in S_A$ of A, there is no dif-
ference between the procedure of constructing the state $F(o, s)$ and (iv) but
placing the signal in the opposite direction.

In this definition, the rectangle region $\{(x, y) \in \mathbb{Z} \times \mathbb{Z} \mid x_{min} \leq x \leq x_{max} \wedge y_{min} \leq y \leq y_{max}\}$ is called the *frame* of F.

Note that if functions G and g are given, the wire-based simulation function F is uniquely determined by the conditions (iv) and (v) of Definition 12. So we regard the pair of G and g as F. In figures, $G(s)$ for a state $s \in S_A$ is shown as an arrangement of cell states and $g(io)$ is shown by an arrow pointing at the cell corresponding to the input/output terminal io.

Simulation of K. We construct a wire-based simulation of the s-automaton K now. Figure 4 shows the state $G_K(0)$ and the coordinate $g_K(io)$ for each terminals $io \in I_K \sqcup O_K$. They give wire-based simulation F_K of the s-automaton K in state $0 \in S_K$. Figure 5 shows the state $F_K(0, 0)$. The states simulating K with a signal in the other input/output terminals are also constructed in the same way.

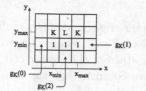

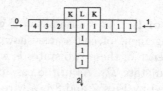

Fig. 4. $G_K(0)$ and g_K.   **Fig. 5.** $F_K(0, 0)$.

Now we confirm that the function F_K determined by Fig. 4 and the conditions of Definition 12 is a wire-based simulation function of K. Since there is no rule in Table 1 that is applicable to a cell in $G_K(0)$, the state $G_K(0)$ cannot be changed. Thus, the condition (i) of Definition 12 is satisfied. Recall that the cells which are not drawn in figures are assumed to be in state $0 \in S_M$. That means the condition (ii) of Definition 12 is satisfied if the frame of F_K is determined by $x_{min}, x_{max}, y_{min}$ and y_{max} in Fig. 4. The condition (iii) of Definition 12 is also satisfied because of the way to interpret the figure. The conditions (iv) and (v) are satisfied because the function F_K is constructed by these conditions. Thus, if the function F_K is a simulation function of K, F_K is a wire-based simulation function.

Figure 6 shows the transitions that can be made if the initial state is $F_K(0, 0)$. Many transitions are presented by $\rightarrow^*$ to save space. These omitted transitions are almost the same as the transitions shown in Fig. 2. Any state t' which satisfies $F_K(0, 0) \rightsquigarrow_{F_K} t'$ appears in the transitions shown in Fig. 6, and any state t' in this transitions satisfies $t' \rightsquigarrow_{F_K} F_K(2, 0)$. A similar argument holds for transition $(1, 0) \rightarrow_K (2, 0)$, so the condition (i) of Definition 2 is satisfied.

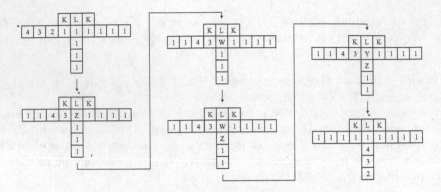

Fig. 6. The transitions from $F_K(0,0)$ to $F_K(2,0)$.

Since there is no transition of M to the states $F_K(0,0)$ or $F_K(1,0)$ and there is no transition of M from the state $F_K(2,0)$, the condition (ii) of Definition 2 is also satisfied.

Therefore the function F_K is a simulation function of K.

Simulation of E. The s-automaton E with the inner state u or d is simulated by the arrangement of cell states shown in Fig. 7 or Fig. 8, respectively. The unique difference of these two states is whether the state of the center cell is in state U_0 or in state D_0. As in the case of K, we can confirm that the function F_E determined by Figs. 7 and 8 is a wire-based simulation of E.

The procedures to confirm that transitions starting from a state on M satisfy conditions of Definition 2 are mechanical but complicated. In practice, we conducted these procedures by using a computer program, which simulates transitions on M and checked the conditions of Definition 2.

Crossing. Since we are dealing with two-dimensional space, it may be difficult to connect two terminals with a wire. We solve this problem by constructing a crossing, which allows two wires to cross each other. The construction of a crossing is shown in Fig. 9.

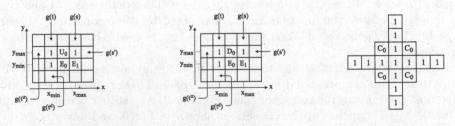

Fig. 7. $G_E(u)$.   **Fig. 8.** $G_E(d)$.   **Fig. 9.** Crossing.

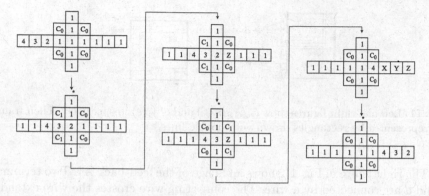

Fig. 10. A signal on a crossing.

Figure 10 shows how a signal progresses across another wire. A wire cell in state 1 is usually changed to state Z when a cell in state 2 is in its neighborhood. However, a wire cell in contact with a cell in state C_1 is not changed because of rule 1 of Table 1. That is the reason why a signal progresses straight at the center of a crossing. These transitions are non-deterministic, but any succeeding state transfers to the state having a signal in the opposite side of the initial state. Thanks to crossings, we can connect terminals freely.

Product and Feed-Back. We have proved that there are wire-based simulation functions of K and E. Next we will explain how to combine them.

Let $A = (I_A, O_A, S_A, \rightarrow_A)$ and $B = (I_B, O_B, S_B, \rightarrow_B)$ be s-automata. Assume that there are wire-based simulation functions F_A of A and F_B of B. Then, we can construct a wire-based simulation $F_{A \otimes B}$ of $A \otimes B$ by the following procedure.

First, we assume that all coordinates corresponding to terminals adjoin the right edge of the frame of the wire-based simulation functions. This assumption is possible because we can extend wires freely. Second, we construct $G_{A \otimes B}(s, s')$ for state $(s, s') \in S_A \times S_B$ by putting $G_A(s)$ and $G_B(s')$ in parallel vertically in such a way that the right sides are aligned. Then, $g_{A \otimes B}$ indicates the coordinates to which the wires have been extended in the first step. The function $F_{A \otimes B}$ determined by these $G_{A \otimes B}$ and $g_{A \otimes B}$ is a wire-based simulation of $A \otimes B$. The left figure of Fig. 11 shows the image of $G_{A \otimes B}(s, s')$.

Let $A = (I_A, O_A, S_A, \rightarrow_A)$, $x \in I_A$ and $y \in O_A$ be an s-automaton, its input terminal and its output terminal, respectively. Assume that there is a wire-based simulation function F_A of A. A wire-based simulation function $F_{A_y^x}$ of the feed-back A_y^x is constructed as follows. First, a wire is extended from $g_A(y)$ to $g_A(x)$. Next, the frame of $F_{A_y^x}$ is installed so that it includes the whole frame of F_A and the wire extended in the first step. Finally, for each terminal $io \in (I_A \backslash \{x\}) \sqcup (O_A \backslash \{y\})$, a wire is extended outward from $g_A(io)$, and the coordinate $g_{A_y^x}(io)$ is determined by the end of the extended wire. If necessary, wires cross each other by using a crossing.

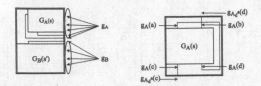

Fig. 11. Left and right figures show $G_{A \otimes B}(s, s')$ and $G_{A_d^a}(s)$, respectively. Their frames are represented by rectangles drawn with double lines.

The right figure of Fig. 11 shows an image of the feed-back A_d^a. Two terminals a and d are connected by a wire. The connecting wire crosses the wire extended from $g_A(b)$, so the crossing must be used at the point.

Infinite Chain. Let A and B be s-automata in Priese System. There exist wire-based simulation functions F_A and F_B. Assume that A and B satisfy the assumptions of Definition 11. A wire-based simulation of the infinite chain made of A and B can be constructed by putting G_A and infinite copies of G_B in sequence and connecting the corresponding terminals to each other.

4   Discussion

This paper has presented an asynchronous non-camouflage CA, which is suitable for implementation by a reaction-diffusion system. The automaton has been obtained by making some changes to the asynchronous totalistic CA presented in [2]. The asynchronous CA presented in this paper is Turing-complete as expected.

The transition function of the asynchronous CA proposed in this paper is represented by less rules than the previous one is; 23 rules of the transition function of the former CA are less than half of 57 rules of that of the latter CA. This is a surprising result because the non-camouflage property is stronger than the outer totalistic property. This result will make it easy to design DNA reactions corresponding to the transition function.

The factor that can be credited for this result is the high expressiveness of the Boolean totalistic form. For example, the transition from cell state 2 to cell state Y is common in both of asynchronous CA. In outer totalistic form, this transition is represented by 7 rules. The number of states of neighboring cells changes depending on where the wire cell is, so a distinct rule corresponding to each situation is required. In our Boolean totalistic form, the transition is represented by just one rule (see rule 5 in Table 1).

We succeeded in constructing an asynchronous CA suitable for implementation by a reaction-diffusion system, but several practical obstacles remain until this implementation can be realized. The proposed cellular automaton has 21 cell states, but it is necessary to reduce the number of cell states to actually be able to implement the CA by a reaction-diffusion system with DNA molecules.

It is also necessary to further simplify the transition rule so that the number of reactions is reduced. Finally, the intrinsic universality of non-camouflage CA, i.e., its Turing-completeness restricted to finite configurations is an interesting open problem.

Acknowledgements. This work was supported by a Grant-in-Aid for Scientific Research on Innovative Areas "Molecular Robotics" (No. 15H00825 and No. 24104005) of The Ministry of Education, Culture, Sports, Science, and Technology, Japan.

References

1. Hagiya, M., Wang, S., Kawamata, I., Murata, S., Isokawa, T., Peper, F., Imai, K.: On DNA-based gellular automata. In: Ibarra, O.H., Kari, L., Kopecki, S. (eds.) UCNC 2014. LNCS, vol. 8553, pp. 177–189. Springer, Cham (2014). doi:10.1007/978-3-319-08123-6_15
2. Isokawa, T., Peper, F., Kawamata, I., Matsui, N., Murata, S., Hagiya, M.: Universal totalistic asynchonous cellular automaton and its possible implementation by DNA. In: Amos, M., Condon, A. (eds.) UCNC 2016. LNCS, vol. 9726, pp. 182–195. Springer, Cham (2016). doi:10.1007/978-3-319-41312-9_15
3. Kawamata, I., Hosoya, T., Takabatake, F., Sugawara, K., Nomura, S., Isokawa, T., Peper, F., Hagiya, M., Murata, S.: Pattern formation and computation by autonomous chemical reaction diffusion model inspired by cellular automata. In: 2016 Fourth International Symposium on Computing and Networking (CANDAR), pp. 215–221. IEEE (2016)
4. Morita, K.: A simple universal logic element and cellular automata for reversible computing. In: Margenstern, M., Rogozhin, Y. (eds.) MCU 2001. LNCS, vol. 2055, pp. 102–113. Springer, Heidelberg (2001). doi:10.1007/3-540-45132-3_6
5. Priese, L.: Automata and concurrency. Theor. Comput. Sci. **25**(3), 221–265 (1983)
6. Wang, S., Imai, K., Hagiya, M.: On the composition of signals in gellular automata. In: 2014 Second International Symposium on Computing and Networking (CANDAR), pp. 499–502. IEEE (2014)

Author Index

Printed in the United States
By Bookmasters